The Naked Spinor

A Rewrite of Clifford Algebra

by

Dennis Morris

Published by: Abane & Right

31-32 Long Row

Port Mulgrave

Saltburn

TS13 5LF

01287 678918

January 2015

Revised July 2015

Revised February 2018

Contents

Contents

Contents

Contents

Contents

References:

We are most grateful to Pertti Lounesto for his elucidation of Clifford algebras in the book:

Clifford Algebras and Spinors 2nd edition

Pertti Lounesto

Cambridge University Press

ISBN: 0-521-00551-5

We are also grateful to Howard Georgi for his book:

Lie Algebras in Particle Physics 2nd edition

Howard Georgi

Westview Press

ISBN: 0-7382-0233-9

We will frequently refer to these books using only the author's surname.

Prerequisites:

This book will be more easily understood by the reader if the reader has some familiarity with higher dimensional rotations and higher dimensional trigonometry in general. Such familiarity can be gained by reading the forerunner to this book which is:

The Physics of Empty Space

by

Dennis Morris

ISBN: 978-1-507707-00-5

2015

The reader might also want to read:

Complex Numbers The Higher Dimensional Forms

by

Dennis Morris

ISBN: 978-1508877499

2007

Chapter 1

What is a Spinor?

"No-one fully understands spinors. Their algebra is formally understood, but their general significance is mysterious. In some sense, they describe the "square root" of geometry and, just as understanding the square root of minus one took centuries, the same might be true of spinors." [1] - Michael Atiyah[2].

Your author believes that he does understand spinors at least as fully as he understands the square root of minus one. Your author is not a particularly brilliant mathematician, and he is certainly not a mathematician of the eminence of Michael Atiyah, but he is a lucky mathematician.

Some ten years ago, your author stumbled upon a change in the notation we use to write complex numbers. Instead of writing a complex number as $a + ib$, your author began to write a complex number as a 2×2 matrix. From nothing more than this change of notation, your author discovered the higher dimensional complex numbers. The higher dimensional complex numbers include the whole of Clifford algebra and, with that, the whole of spinor theory[3]. The change of notation simplifies and generalises Clifford algebra and spinor theory. Just a small change in notation, nothing more, has allowed your author to accomplish in ten years what one of the greatest mathematicians of the 20th century thought might take centuries. In this book, your author hopes he lays bare the nature and general significance of spinors.

---

[1] Quoted in: "The Strangest Man. The Hidden Life of Paul Dirac, Quantum Genius", Graham Farmelo, Faber & Faber 2009. ISBN: 978-0-571-22286-5 pg 430
[2] Michael Atiyah is a mathematician of world-wide repute.
[3] They do contain the whole of spinor theory, but most of it is not yet known.

The cacophony:

There is much confusion within the literature regarding spinors. They are defined, or at least described, in a dozen different ways; not all definitions match exactly, and not all descriptions look the same. Nor do spinors have their own particular place within mathematics and physics. They are like a bad smell with which no-one wants to be associated.

To a physicist, a spinor is a convenient pair, or two pairs, of complex numbers and nothing more[4]. The physicist uses spinors without much regard to what they are or how they relate to other areas of mathematics. Spinors give the physicist the right answers, and that is all the physicist needs. Nonetheless, the physicist knows that spinors have 'strange properties', like double cover, and are associated with strange phenomena like the intrinsic spin of an electron. These 'strange properties' give them a sense of the mystical as expressed by Atiyah's words *"No-one fully understands... their general significance is mysterious"* quoted above.

To many mathematicians, but not all mathematicians, a spinor is an element with peculiar properties within some Clifford algebras. To a Clifford algebraist, a spinor is a distraction from the central theme of Clifford algebras, and he wonders what spinors are doing in Clifford algebra. Clifford algebraists will often define and describe spinors in many seemingly very different ways. One is driven to wonder if they really know that of which they speak. This situation is not helped by the alleged existence of 6-dimensional spinors although there are no 6-dimensional Clifford algebras or 6-dimensional even sub-algebras of Clifford algebras. It seems that the history of physics, Clifford algebras, and spinors, has contributed to this cacophonous confusion.

The history of spinors:

The Clifford algebras were invented by William Kingdon Clifford (1845-1879) in 1876; his work was published posthumously in 1882.

---

[4] We will see how this view emerges from the division algebras as an 'expectation spinor'; this is like an expectation value in quantum physics.

The Clifford algebras were also invented independently by Rudolf Lipschitz (1832-1903) in 1880 who acknowledged Clifford's priority in 1886.

Clifford algebra was seen then as being a mathematics of empty space, $\mathbb{R}^n$, and there was not a spinor in sight. Even today, Clifford algebra is still seen as the mathematics of empty space and spinors are pushed into a dark corner of Clifford algebra.

Spinors were discovered independently of Clifford algebra by Elie Cartan in 1913. Cartan was working in $\mathbb{C}^n$ types of linear space, and he discovered that $\mathbb{C}^3$ space contains a $\mathbb{C}^2$ sub-space with 'strange' properties. Only later did Cartan discover that spinors were connected to Clifford algebras. Hence, spinors are often seen, as being an 'add-on' to Clifford algebra. The position of spinors within Clifford algebra is not helped by the seeming appearance of spinors in only some of the Clifford algebras.

Physicists began to use spinors in 1928 because they worked, but the physicists paid little concern to the mathematical position or heritage of spinors. Physicists habitually care not for Clifford algebra, and, although they are aware of it, they do not use it.

The stance of this book:
With over a dozen different definitions of spinors floating around, the reader might think that the last thing she needs is yet one more definition of spinors. Sorry, we are going to give that unwanted definition. Explaining the definition we give is what the remainder of this book is about. The definition we give is much simpler than any previous definition; it also mutes the cacophony of multiple definitions because it includes all other definitions and descriptions. It is a comprehensive and general definition within which we can easily make sub-definitions if we choose. The mathematics associated with this new definition is nothing more than division algebras and is much simpler than the mathematics associated with previously given definitions, but it is mathematics with which the

reader might be unfamiliar. Division algebra mathematics is little more than matrices and elementary finite group theory, and so, although perhaps unfamiliar to the reader, it is easy to learn and use.

To aid clarification, we will give two examples of a spinor along with the definition.

Definition of a spinor:
A spinor is a unit length element of a division algebra. This is a matrix with determinant unity.

Clarification of the spinor definition:
There are many division algebras. Division algebras are just types of numbers like the complex numbers, $\mathbb{C}$, or the quaternions, $\mathbb{H}$. Each division algebra has its own set of spinors (unit length elements).

All division algebras can be written in their polar form which is a real number, the radial variable, multiplied by a rotation matrix. The given definition means that a spinor is an element of the rotation matrix of a division algebra.

Examples of spinors:
We see that, by our definition, a unit length single complex number is a spinor[5]. We have the complex numbers in polar form:

$$\begin{bmatrix} r & 0 \\ 0 & r \end{bmatrix} \begin{bmatrix} \cos\theta & \sin\theta \\ -\sin\theta & \cos\theta \end{bmatrix} \tag{1.1}$$

By our definition, the set of spinors in this division algebra is the rotation matrix for all values of θ.

---

[5] Some authors count a single complex number as a spinor. Some authors do not count a single complex number as a spinor. Some authors do both. This is one of the ambiguities within the Clifford algebra approach to spinors.

We see that, by our definition, a unit length quaternion is a spinor[6]. We have the quaternion rotation matrix (the angular part of the polar form of the quaternions):

$$
\mathbb{H}_{Rot} =
\begin{bmatrix}
\cos(\lambda) & \dfrac{b}{\lambda}\sin(\lambda) & \dfrac{c}{\lambda}\sin(\lambda) & \dfrac{d}{\lambda}\sin(\lambda) \\[2.5ex]
-\dfrac{b}{\lambda}\sin(\lambda) & \cos(\lambda) & -\dfrac{d}{\lambda}\sin(\lambda) & \dfrac{c}{\lambda}\sin(\lambda) \\[2.5ex]
-\dfrac{c}{\lambda}\sin(\lambda) & \dfrac{d}{\lambda}\sin(\lambda) & \cos(\lambda) & -\dfrac{b}{\lambda}\sin(\lambda) \\[2.5ex]
-\dfrac{d}{\lambda}\sin(\lambda) & -\dfrac{c}{\lambda}\sin(\lambda) & \dfrac{b}{\lambda}\sin(\lambda) & \cos(\lambda)
\end{bmatrix}
\tag{1.2}
$$

$$
\lambda = \sqrt{b^2 + c^2 + d^2}
$$

By our definition, the set of spinors in this division algebra is the rotation matrix for all values of $\{b,c,d\}$.

To a Clifford algebraist, the above quaternion rotation matrix, (1.2), is (isomorphic to) a spin group, $Spin(3)$, and it is the elements of the spin group that are spinors. The reader can think of a spin group as being a division algebra rotation matrix like the quaternions above, (1.2); it has to be the rotation matrix of a division algebra; the 2-dimensional rotation matrices in 3-dimensional space of $SO(3)$ are not spin groups. No division algebra has a polar form with a $SO(3)$ rotation matrix.

---

[6] All authors agree that, up to isomorphism, a unit length quaternion is a spinor, but few state this fact. Further, some authors take the whole of the quaternion algebra, elements of any length, to be spinors rather than restrict the definition to only unit length quaternions. Other authors take the view that quaternions do not exist within Clifford algebras but that some Clifford algebras have even sub-algebras that are isomorphic to the quaternions and that it is the elements of these even sub-algebras that are spinors. You see the cacophony.

Discussion of the spinor definition:

Some Clifford algebraists include the unit length complex numbers, $\mathbb{C}_{Rot}$, in their definition of spinors, and some Clifford algebraists do not include the unit length complex numbers in their definition of spinors. For some Clifford algebraists, a 'spin group' (division algebra rotation matrix to us) must be a double cover of an orthogonal group, $SO(p,q)$, and a single complex number is not such a double cover. Clearly, our definition includes the unit length complex numbers as spinors.

The quaternion rotation matrix is a double cover of $SO(3)$. Let us set two of the quaternion variables to zero and thereby reduce the quaternion rotation, (1.2), to a 2-dimensional rotation[7] in quaternion space:

$$
\begin{bmatrix}
\cos\left(\sqrt{b^2}\right) & \sin\left(\sqrt{b^2}\right) & 0 & 0 \\
-\sin\left(\sqrt{b^2}\right) & \cos\left(\sqrt{b^2}\right) & 0 & 0 \\
0 & 0 & \cos\left(\sqrt{b^2}\right) & -\sin\left(\sqrt{b^2}\right) \\
0 & 0 & \sin\left(\sqrt{b^2}\right) & \cos\left(\sqrt{b^2}\right)
\end{bmatrix}
\tag{1.3}
$$

If we put $\theta > 0$ into the complex number rotation matrix, (1.1), we get a clockwise rotation. If we put $\theta < 0$ into the complex number rotation matrix, (1.1), we get an anti-clockwise rotation. Simple, we are accustomed to this type of rotation. However, if we put $b > 0$ into the quaternion rotation matrix, (1.3), we get both a clockwise rotation and an anti-clockwise rotation within the same rotation – the minus signs in (1.3) are oppositely distributed. Also, if we put $b < 0$ into the quaternion rotation matrix, (1.3), we still also get both a clockwise rotation and an anti-clockwise rotation. Your author asserts that this doubling up of rotations such that we get both a clockwise rotation

---

[7] This is a 4-dimensional 2-dimensional rotation. This is different from the type of 2-dimensional rotation to which we are accustomed. It is a double cover rotation for a start, but, perhaps more importantly, it is not rotation about an axis – more later.

and anti-clockwise rotation in the same rotation is what is usually called double cover.

Let us put $b = \dfrac{\pi}{2}$ into (1.3). We get two rotations, clockwise and anti-clockwise, for a single angle. By the time we have rotated through 2π, we have had 4π's worth of normal rotation. Furthermore, at any point other than $b = n\pi$, we can swap the sign of the $\sin\left(\sqrt{b^2}\right)$ terms by simply choosing the opposite sign of the square root inside the trigonometric function.

Conventionally, double cover is a mysterious bit of spinor mathematics and is often expressed in unclear and complicated mathematics. Above, (1.3), we see that double cover is no more than the true nature of quaternion rotation[8].

A common definition of spinors is that they are a double cover of the $SO(p,q)$ special orthogonal groups. If we were to accept this definition, then we would have spinors in only the division algebras that derive from the $C_2 \times C_2 \times...$ finite groups for these are the algebras that have double covers in 2^n dimensions where n is the number of copies of the C_2 group which are crossed together. We will see in due course that the whole of Clifford algebra is contained within the division algebras that derive from the $C_2 \times C_2 \times...$ finite groups; this is the connection between Clifford algebra and spinors.

Multiple arguments in the trigonometric functions:
We note that the trigonometric functions in the 2-dimensional rotation matrix, (1.1), accept only one argument, θ, whereas the quaternion trigonometric functions in (1.2) accept three arguments, $\{b,c,d\}$. Quaternion rotation in 4-dimensional quaternion space is

---

[8] It is a little more complicated than this because quaternion rotations are not commutative, but this complication is in the minds of mathematicians not in the mathematics – more later.

accomplished by one 4×4 rotation matrix. Rotation in our 4-dimensional space-time is accomplished by six 2×2 rotation matrices. We see that these two types of rotation are very different. To the physicist, the two types of rotation correspond to the two types of angular momentum, orbital angular momentum and intrinsic spin angular momentum. In general, division algebras of dimension higher than two have trigonometric functions that accept more than one argument. We could correctly think of such multi-angle rotation as spinor rotation, but, politely, we include the 2-dimensional division algebra rotations to be neat. To exclude the 2-dimensional rotations would be messy and would gain nothing.

No rotation about an axis:

The spinor rotations are not rotation about an axis. This is most strange to we inhabitants of 4-dimensional space-time. Rotation in the complex plane, (1.1), is not rotation about an axis because the complex plane is 2-dimensional. There is no third dimension sticking out of the complex plane forming an axis of rotation. Consider an element of the group $SO(3)$ which are the 2-dimensional rotations in three spatial dimensions to which we are accustomed:

$$SO(3)_1 = \begin{bmatrix} \cos\theta & \sin\theta & 0 \\ -\sin\theta & \cos\theta & 0 \\ 0 & 0 & 1 \end{bmatrix} \tag{1.4}$$

The eigenvalues and eigenvectors of this rotation matrix include:

$$1 \ \& \ \begin{bmatrix} 0 \\ 0 \\ 1 \end{bmatrix} \tag{1.5}$$

This is an eigenvector which is unchanged by the rotation (it is independent of θ). This is the spatial direction, unchanged by rotation, which we call the rotation axis. If we take the eigenvalues and eigenvectors of the 2-dimensional rotation matrix (1.1), we do not find an eigenvector that is independent of the rotation angle, θ. Similarly, there is no eigenvector of the 4-dimensional 2-dimensional quaternion rotation matrix (1.3) which is independent of the angle. Spinor rotation is not rotation about an axis. We can see why electron spin is 'weird' to our eyes.

By the way, 2-dimensional rotation in our 4-dimensional space-time is rotation about two axes – there are two eigenvectors which are independent of the angle.

Non-commutative rotations:

The Clifford algebras are all non-commutative. Therefore, all traditional spinors are non-commutative. The definition we have given above allows commutative rotations, but the commutative rotations, are, as far as we know in general, not double covers of $SO(p,q)$ type groups. We will look at a quaternion rotation as an example of a double cover rotation. The quaternions are non-commutative.

A vector within a division algebra is an element of that algebra. As such, we write a vector as a square matrix rather than as a column of real numbers; for example, the point $(1,2)$ in the complex plane would conventionally be written as a column of two numbers and called a vector; we write the point $(1,2)$ as a square matrix and call it a vector:

$$\begin{bmatrix} 1 \\ 2 \end{bmatrix} \equiv \begin{bmatrix} 1 & 2 \\ -2 & 1 \end{bmatrix} \tag{1.6}$$

Similarly, a vector in quaternion space is written as a 4×4 matrix rather than as a column of four numbers.

If we multiply a quaternion vector (square matrix) on the right by a quaternion rotation matrix, we get a different result from the result we get if we multiply a quaternion vector on the left by a quaternion rotation matrix – quaternions are non-commutative, after all is said and done:

$$\mathbb{H}_{Rot}\mathcal{Q}_{Vector} \neq \mathcal{Q}_{Vector}\mathbb{H}_{Rot} \tag{1.7}$$

These are rotations through the same angle, but we have two different rotations. We are accustomed to associating only one rotation with one angle. We could agree to always rotate on the left (or right) of the vector, but the usual approach taken by Clifford algebraists is to rotate on both the left and the right:

$$\mathbb{H}_{Rot}\mathcal{Q}_{Vector}\mathbb{H}_{Rot} \tag{1.8}$$

This does give only one rotation for one angle, but, because we have rotated twice, we rotate through twice the angle in the rotation matrix. We put 360^0 into the rotation matrices, and we get 720^0 worth of rotation. This is not really double cover, but it is what Clifford algebraists often call double cover.

Other possible definitions:

We could have defined spinors to be the whole of the division algebras rather than just the unit length elements of the division algebras. This would be an entirely sensible definition, and since we get the whole of a division algebra by simply multiplying the rotation matrix by a real number, in some ways we have defined a spinor to be any element of a division algebra. We prefer the unit length element definition because spinors are associated with rotation in physics. Ultimately, the choice is arbitrary.

We could have defined spinors to be unit length elements of only the division algebras that derive from the $C_2 \times C_2 \times...$ groups. This would correspond in lower dimensional spaces to defining them as double covers of some of the $SO(p,q)$ groups.

The general definition we have preferred allows us to classify spinors into different sets such as double covering spinors or the spinors from cyclic groups. We therefore include the less general definitions within the general definition as sub-definitions.

Simply stated, the general definition of a spinor is that a spinor is an element of a rotation matrix of a division algebra.

What is to be done:
We are about to lay before the reader the whole of conventional spinor theory[9]. Spinor theory is the same as the theory of higher dimensional division algebras (higher dimensional complex numbers) except that, in both cases, we include the lower dimensional division algebras to be neat. To do this, we need the reader to be familiar with the higher dimensional division algebras and how they derive from the finite groups. Part of this book will explain that, but the reader will find easier introductions in the books listed in the prerequisites at the start of this book.

The whole of Clifford algebra is within the division algebras that derive from the $C_2 \times C_2 \times ...$ finite groups. Since Clifford algebras and spinors are historically intertwined, we need to expound Clifford algebra and relate it to these division algebras so that the reader might understand the relationship between Clifford algebras and spinors. This is effectively to rewrite the whole of Clifford algebra. Part of this book is concerned with presenting the conventional exposition of Clifford algebras and then rewriting Clifford algebra as division algebras.

Summary:
A spinor is an element of a rotation matrix of a division algebra.

---

[9] Actually, there are huge gaps in humankind's understanding of spinor theory, but we are able to present a rewrite of the whole of what is currently known.

Historical note:

In 1845, Cayley published[10] the rotation formula:

$$\vec{y} = a\vec{x}a^{-1} \tag{1.9}$$

Wherein $\{\vec{x}, \vec{y}\}$ are vectors (points) and a is a quaternion of the form:

$$a = \cos\frac{\lambda}{2} + \frac{\vec{a}}{\lambda}\sin\frac{\lambda}{2} \tag{1.10}$$

Wherein $\vec{a}$ is the 'vector part' of the quaternion $\{0 + ib + jc + kd\}$ and $\lambda = |\vec{a}| = \sqrt{b^2 + c^2 + d^2}$. When the notation is untangled, this is a quaternion rotation matrix with quaternion trigonometric functions with a half-angle rather than an angle. Cayley credited Hamilton with the discovery.

---

[10] A. Cayley: On certain results relating to quaternions: Phil Mag (3) 26 (1845) pg 141-145. Lounesto: Page 321

Chapter 2

First Steps

In this book, we will rewrite Clifford algebra and show it to be no more than, and somewhat less than, the non-commutative $C_2 \times C_2 \times ...$ division algebras[11]. We are going to produce the complete theory of spinors. We will see that the $C_2 \times C_2 \times ...$ division algebras, and hence Clifford algebra, are only a part of that complete theory.

At another level, this book is really about empty space; it is about two different types of empty space. These two different types of space are spinor space and $\mathbb{R}^n$ space. These two different types of space are division algebra space and Riemannian type space, and this book is about these spaces and how they are related.

It seems to your author that conventional Clifford algebras attempt to simultaneously exist in both these types of space, and, in so doing, in your author's opinion, the foundations of Clifford algebra are questionable. None-the-less, Clifford algebras are highly developed and they contain a great deal of useful mathematics; one does not want to throw away all this knowledge. If we can rewrite the foundations of Clifford algebra, we can then sit upon the shoulders of our forebears and re-interpret their contributions to Clifford algebra in our own rewrite.

The two types of geometric space:
There are two types of geometric space. The first of these types is the n-dimensional division algebra spaces like the complex plane, or the

---

[11] These are the division algebras that derive from the $C_2 \times C_2 \times ...$ groups. Although these are commutative groups, they hold non-commutative division algebras.

quaternions, or the A_3 algebras[12]. These n-dimensional division algebra spaces have one real axis and $(n-1)$ imaginary axes. These are the spinor spaces. N-dimensional division algebra spaces derive from the finite groups of order n and are division algebras whose polar form includes a $n \times n$ rotation matrix containing n different n-dimensional trigonometric functions[13] (n of each type); an example is the 4×4 quaternion rotation matrix, (1.2). Rotations in these spaces are true n-dimensional rotations rather than the compilations of 2-dimensional rotations we have in the $\mathbb{R}^n$ spaces. Rotations in division algebra spaces are rotations in spinor spaces.

Division algebra spaces contain p-dimensional sub-spaces only if the finite group from which they are derived contains sub-groups of order p; for example, there are no 2-dimensional sub-spaces within the 3-dimensional division algebra spaces because the finite group C_3 does not contain a C_2 sub-group. Contradistinctly to division algebra spaces, $\mathbb{R}^n$ spaces contain sub-spaces of any dimension less than $n \geq 0$. We can simply 'wrench off' a few axes. One cannot form the finite group C_2 by simply 'wrenching off' an element of the finite group C_3.

Division algebra spaces come in algebraically isomorphic sets; the members of a set differ from each other by only a change of variable or change of basis; for example, there are two quaternion algebras and there are six A_3 algebras.

The second type of geometric space is the n-dimensional $\mathbb{R}^n$ spaces. These spaces are derived by superimposing all the elements of a set of isomorphic division algebra spaces[14]. The 4-dimensional space-time in which we sit is such a space; it is formed by the superimposition of the six A_3 algebras. When we superimpose all elements of a set of

---

[12] These will be introduced later.
[13] The reader will be introduced to the higher dimensional trigonometric functions later in this book.
[14] We'll do this later.

algebraically isomorphic division algebra spaces written in different bases to make a $\mathbb{R}^n$ space, we break the division algebras, and thus we break the symmetries (e.g.: rotation as expressed by the rotation matrix in the polar form of the algebra) in those algebras. We opine that such breaking of algebras is associated with the emergence of classical physics from quantum physics. Since algebraic multiplication[15] can exist in only division algebras, there is no algebraic multiplication in $\mathbb{R}^n$ spaces and so there are no imaginary axes in $\mathbb{R}^n$ spaces. The nature of commutation relations in $\mathbb{R}^n$ spaces is also different from the nature of commutation relations in division algebra spaces. Commutation relations in $\mathbb{R}^n$ spaces are relations between different $n \times n$ rotation matrices whereas commutation relations within a division algebra are relations between individual imaginary variables within a single rotation matrix.

Because there is only one division algebra in each set of 2-dimensional division algebra spaces, (the two sets, $\mathbb{C}\,\&\,\mathbb{S}$), the 2-dimensional algebras are not broken by the superimposition operation, and so the symmetries of these spaces are not broken and the 2-dimensional rotations survive the superimposition operation. This is why we see 2-dimensional rotations in our space-time. Higher dimensional rotations[16] do not survive the superimposition operation[17], and so rotation in $\mathbb{R}^n$ spaces can never be other than 2-dimensional and is accomplished by several different rotation matrices that contain 2-dimensional trigonometric functions and some zeros and ones.

In $\mathbb{R}^n$ spaces, once a co-ordinate system is set, any point in the 'spherical shell' surrounding the origin can be reached from a different point in the 'spherical shell' only by a sequence of 2-dimensional rotations. In division algebra spaces, once a co-ordinate system is set, any point in the 'spherical shell' surrounding the origin can be reached by a single n-dimensional rotation. We see that rotations in $\mathbb{R}^n$ spaces

---

[15] This is multiplication in the pedantic mathematical sense – proper multiplication.

[16] We will introduce higher dimensional rotations later; the quaternion rotation matrix is an example of a 4-dimensional rotation.

[17] In three dimensions, there survive two such 3-dimensional rotations, but these spaces do not have Riemannian distance functions.

differ from rotations in division algebra spaces. The difference is related to the fact that rotations in $\mathbb{R}^n$ spaces are not simply connected whereas rotations in division algebra spaces are simply connected. The astute reader might know that spinor spaces are simply connected.

Clifford Algebras:

Conventionally, the Clifford algebras confuse both types of geometric space. The Clifford algebras are presented as 2^n dimensional algebras over the reals[18]; this means that the basis vectors are in $\mathbb{R}^n$ space. However, the basis vectors and basis multi-vectors are the square roots of plus unity or minus unity[19], which means they are in division algebra spaces; only the scalar axis of a Clifford algebra is truly real. An example of this confusion is the mixing of axes[20]:

$$Cl_{2,0} = Cl_{2,0}^+ \oplus Cl_{2,0}^- = \mathbb{C} \oplus \mathbb{R}^2 \tag{2.1}$$

Rotation groups in Clifford algebras are called spin groups. They are double covers of the rotation groups, $SO(p,q)$, in $\mathbb{R}^n$ spaces. In this way, Clifford algebras are clearly not in $\mathbb{R}^n$ space. Spin groups are often isomorphic to special unitary groups; we have:

$$spin(3) \cong SU(2) \tag{2.2}$$

There is another point of potential confusion. Clifford algebras are algebraically isomorphic to matrix algebras; we have:

$$SU(2) = \left\{ U \in Mat(2,\mathbb{C}) \ : \ \det(U) = 1, \ U^\dagger U = 1 \right\} \tag{2.3}$$

and[21]:

---

[18] Lounesto: page 14
[19] Lounesto: page 8
[20] Lounesto: page 28
[21] Lounesto: pages 14, 54

$$Cl_{2,0} \cong Mat(2, \mathbb{R})$$
$$Cl_{3,0} \cong Mat(2, \mathbb{C}) \tag{2.4}$$

$$Cl_{4,0} \cong Mat(2, \mathbb{H})$$
$$Cl_{1,3} \cong Mat(2, \mathbb{H}) \tag{2.5}$$
$$Cl_{3,1} \cong Mat(4, \mathbb{R})$$

The first of the above, (2.4), are the 2×2 matrices with real elements. The second of the above, (2.4), are the 2×2 matrices with complex elements. The third and fourth of the above, (2.5), are the 2×2 matrices with quaternion elements, and the fifth of the above, (2.5), are the 4×4 matrices with real elements.

Your author's view:
In this work, your author takes the view that the Clifford algebras are wrongly perceived. Your author takes the view that the Clifford algebras are all division algebras that derive from the finite groups $C_2 \times C_2 \times...$ This means that they have one real axis and $(n-1)$ imaginary axes. This is the view that will be expounded in this book. This view will initially be unpopular with conventional Clifford algebraists who love the geometric interpretation of Clifford algebra, but your author hopes that they will change this view when they come to understand this rewrite of Clifford algebra.

To expound this view, we need to familiarise ourselves first with Clifford algebras and then with division algebras. That is the purpose of the next few chapters.

A blast from the past:
Spinors were first discovered by the French mathematician Elie Cartan in the first part of the twentieth century. It ought to be noted that, at the very start of spinor theory, a step was taken in the direction

preferred by your author; well, not quite in that direction, by Elie Cartan. We quote a theorem from his initial work on spinors [22]:

"THEOREM: Any matrix of degree 2^v can be regarded, in one and only one way, as the sum of a scalar, a vector, a bi-vector, ..., and a v -vector."

We will be taking each element of a Clifford algebra and writing it as a permutation matrix variable within a matrix. Therefore our matrices will be the same size as the dimension of the Clifford algebra; the dimensions of Clifford algebras are 2^n. Cartan is a long and convoluted way from what we will do, but the idea of putting all elements of a Clifford algebra into the same matrix and thereby ignoring the differences between vectors and bi-vectors is clear.

---

[22] Elie Cartan: The Theory of Spinors: page 86 : Dover Publications: ISBN: 0-486-64070-1

Chapter 3

An Outline of Clifford Algebra

We begin by familiarising the reader with the conventional presentation of Clifford algebra.

When Clifford first formulated the algebras named after him, the mathematical concept of vectors had only recently been invented, and the vector dot product and vector cross product were still unknown. Although vectors could be added, there was no established way to multiply them together. Clifford sought an algebra of vectors which included a way to multiply two vectors together. He wanted a vector product that satisfied the 'sensible' conditions of multiplicative associativity and multiplicative distributivity over addition. Clifford believed that he was working in $\mathbb{R}^n$, and he was aware that, in more than two dimensions, he would have to dispense with multiplicative commutativity to create such a vector algebra within $\mathbb{R}^n$ [23].

In creating such an algebra, Clifford added the concepts of bi-vector and multi-vectors in general to the concept of vector[24]. A vector is associated with an oriented 1-dimensional length, and so Clifford algebraists associated a bi-vector with an oriented area, and a tri-vector with an oriented volume etc.. The orientation is associated with a direction of rotation in the 2-dimensional plane, clockwise or anti-clockwise. Clifford believed he had an algebra that described aspects of *n*-dimensional space; of course, to Clifford, there was only one type of *n*-dimensional space and that was $\mathbb{R}^n$.

---

[23] There are types of division algebra space that derive from the cyclic groups, the C_n spaces, which are commutative in all dimensions.

[24] Trying to put ourselves in the place of the mathematicians who had just discovered these new things called vectors, we too might have poked around to see if there were any other new things to be found. Bi-vectors and tri-vectors are just more new things.

An important aspect of a Clifford algebra is that it takes its basis elements, written as basis vectors, $\vec{e}_i$, to be the square roots of plus or minus unity. For example, in the Clifford algebra $Cl_{2,0}$, we have:

$$1^2 = 1, \quad \vec{e}_1^{\,2} = 1, \quad \vec{e}_2^{\,2} = 1, \quad \vec{e}_{12}^{\,2} = -1 \qquad (3.1)$$

We see here a fundamental confusion within conventional Clifford algebra; the conventional view is that Clifford algebra is an algebra of $\mathbb{R}^n$ space and yet we have the imaginary square roots of ± 1 involved at a very fundamental level.

These days, following the work of David Hestenes and others, Clifford algebras are sometimes known as geometric algebras. Associated with these geometric algebras, there is an extension of Clifford algebras to include differentiation; this extension is known as geometric calculus.

Clifford algebra compared with division algebras:
Division algebras are types of number. Examples are the real numbers, $\mathbb{R}$, the complex numbers, $\mathbb{C}$, or the multiplicatively non-commutative quaternions, $\mathbb{H}$; there are many other types of division algebras. A set of mathematical objects is a division algebra if it satisfies the thirteen division algebra axioms. Multiplicative commutation is not one of the division algebra axioms[25].

The Clifford algebras satisfy all the axioms of a division algebra except for the requirement that every element of the algebra has a multiplicative inverse and an absence of zero divisors[26]. Most elements of a Clifford algebra do have multiplicative inverses, but not all of them have multiplicative inverses, and, essentially, this, and the existence of zero divisors, is the only failing that prevents a Clifford algebra from being a division algebra.

---

[25] A multiplicatively commutative division algebra is an algebraic field.
[26] Zero divisors are two non-zero elements of an algebra whose product is zero. An easy example is two matrices which each have a single non-zero element in different positions.

Aside:

We give an example of zero divisors. In the Clifford algebra $Cl_{0,3}$ we have:

$$\left(\overrightarrow{e_{123}}\right)^2 = 1$$
$$\left(1+\overrightarrow{e_{123}}\right)\left(1-\overrightarrow{e_{123}}\right) = 1-\left(\overrightarrow{e_{123}}\right)^2 = 1-1 = 0$$

(3.2)

Obviously, any Clifford algebra that has a square root of plus unity will contain zero divisors.

A matrix example is:

$$\begin{bmatrix} 1 & 1 \\ 1 & 1 \end{bmatrix}\begin{bmatrix} 1 & -1 \\ -1 & 1 \end{bmatrix} = \begin{bmatrix} 0 & 0 \\ 0 & 0 \end{bmatrix}$$

(3.3)

Wherein we have taken:

$$1 \equiv \begin{bmatrix} 1 & 0 \\ 0 & 1 \end{bmatrix}, \quad \overrightarrow{e_{123}} \equiv \begin{bmatrix} 0 & 1 \\ 1 & 0 \end{bmatrix}$$

(3.4)

Of course most division algebras are multiplicatively commutative and thereby are algebraic fields, but there are non-commutative division algebras (quaternions for example). Many division algebras (the A_3 algebras or the hyperbolic complex numbers are examples) have elements without a multiplicative inverse in the Cartesian form and become division algebras in only the polar form; we take the exponential of the Cartesian form to eliminate the singular matrices and the zero divisors. The Clifford algebras match this property, but taking the exponential of a Clifford algebra to avoid this failing is just not done because it destroys the geometric interpretation of Clifford algebras. If it were done, there would be no absence of multiplicative inverses and no zero divisors and the polar forms of the Clifford algebras thus generated would be division algebras without the conventional geometric interpretation of the algebra.

The algebraic essence of the Clifford algebras:

Although Clifford thought geometrically, and most Clifford algebraists think similarly so, the algebraic essence of the Clifford algebras does not need the concepts of vector, bi-vector, or any concept of geometric space. Howard Georgi[27] defines a Clifford algebra as a set of operators, Γ_i, such that:

$$\{\Gamma_j, \Gamma_k\} = \Gamma_j\Gamma_k + \Gamma_k\Gamma_j = 2\delta_{jk} \tag{3.5}$$

Note that $\delta_{jk} = 0$ if $j \neq k$ & $\delta_{jk} = 1$ if $j = k$. Note that the concept of operators is also not essential to the algebraic essence of the Clifford algebras. Georgi's definition, (3.5), effectively says that $\Gamma_j\Gamma_k = \sqrt{+1}$ if $j = k$ and $\Gamma_j\Gamma_k = \sqrt{-1}$ if $j \neq k$. Products of these elements will be unity, or minus unity, or square roots of plus unity, or square roots of minus unity. Georgi's definition can be extended to include the cases where some or all of the $\Gamma_j\Gamma_k = \sqrt{-1}$ if $j = k$; doing this changes some of the other relations.

What is a Clifford algebra?:

We assert that it is of the algebraic essence of a Clifford algebra that it is concerned with elements that are the square roots of plus unity and elements that are the square roots of minus unity. We will soon argue that the algebraic essence of a *n*-dimensional Clifford algebra is that it is a 2^n dimensional algebra with one real variable and $(2^n - 1)$ imaginary variables that are square roots of plus unity or square roots of minus unity. In other words, we are dealing with something like the higher dimensional forms of the complex numbers – only something like; the complex numbers are a division algebra; the Clifford algebras in general (there are some exceptions) do not have all the properties necessary to be a division algebra; for example, many have zero divisors.

---

[27] Howard Georgi: page 270.

Historically, mathematicians were reluctant to see the Clifford algebras as having one real variable and several imaginary variables because it was believed to have been established that there could be no higher dimensional complex numbers other than the ones already known $\{\mathbb{C}, \mathbb{H}\}$. However, the purported proofs of this assertion are subtle[28]. They assume the 2-dimensional complex numbers must be a sub-algebra of any higher dimensional division algebra and they ignore the existence of the hyperbolic complex numbers[29]; they do not actually prove the stated assertion. The proofs show only that there can be no higher dimensional complex numbers that have a Euclidean quadratic norm with an all positive signature. In this, the proofs are correct, but there are an infinite number of different types of higher dimensional complex numbers that do not have a Euclidean quadratic norm with an all positive signature.

The traditional approach to Clifford algebra:
It is traditional and more usual to see the objects within a Clifford algebra, those objects that Georgi, (3.5), calls operators, as basis vectors in $\mathbb{R}^n$. The Clifford algebra is then given as the products of these basis vectors. The square of a basis vector is either unity or minus unity, and, importantly, the product of two different basis vectors is anti-commutative. In the $Cl_{2,0}$[30] Clifford algebra, those properties are:

$$\vec{e_1}\vec{e_1} = 1 \qquad \vec{e_2}\vec{e_2} = 1$$
$$\vec{e_1}\vec{e_2} = -\vec{e_2}\vec{e_1} \qquad\qquad (3.6)$$

---

[28] There are two proofs, the Hurwitz theorem (1923) and the Frobenius theorem (1878). Lagrange also offered a proof in the 3-dimensional case (circa 1830).

[29] The hyperbolic complex numbers were discovered by Cockle in 1848.

[30] Different Clifford algebras are written as $Cl_{p,m}$. The p is the number of basis vectors that are square roots of plus one; the m is the number of basis vectors that are the square roots of minus one. The reader will often see the $Cl_{2,0}$ algebra written as Cl_2. We include the comma and the zero to avoid confusing ourselves.

These relations of just two basis vectors are sufficient to determine the entire nature of the 4-dimensional Clifford algebra $Cl_{2,0}$. The $\vec{e_1}\vec{e_2}$ is a product of two vectors and is called a basis bi-vector. Squaring this basis bi-vector, and taking account of its non-commutative nature, we have:

$$\left(\vec{e_1}\vec{e_2}\right)\left(\vec{e_1}\vec{e_2}\right) = -\vec{e_1}\left(\vec{e_2}\vec{e_2}\right)\vec{e_1} = -\vec{e_1}\vec{e_1} = -1 \qquad (3.7)$$

The above relations imply:

$$\vec{e_1} = \sqrt{+1}, \qquad \vec{e_2} = \sqrt{+1}, \qquad \vec{e_1}\vec{e_2} = \sqrt{-1} \qquad (3.8)$$

We see that the product of a basis vector with itself or a basis bi-vector with itself produces a scalar (real number). The two basis vectors with the bi-vector and the scalar are the basis of the Clifford algebra denoted $Cl_{2,0}$. That is, $Cl_{2,0}$ is a 4-dimensional Clifford algebra with the basis elements:

$$
\begin{array}{ll}
1 & \text{real scalar} \\
\vec{e_1} = \sqrt{+1} & \text{basis vector} \\
\vec{e_2} = \sqrt{+1} & \text{basis vector} \\
\vec{e_1}\vec{e_2} = \sqrt{-1} & \text{basis bi-vector}
\end{array} \qquad (3.9)
$$

The relations (3.6) mean that a product of more than two basis vectors is one of the other basis elements.

For example:

$$\vec{e_1}\vec{e_2}\vec{e_1} = -\vec{e_1}\vec{e_1}\vec{e_2} = -\vec{e_2} \qquad (3.10)$$

We see the algebra is closed under multiplication. An arbitrary element of this algebra, $Cl_{2,0}$ is:

$$u = u_0 + u_1\vec{e_1} + u_2\vec{e_2} + u_{12}\vec{e_{12}} \qquad (3.11)$$

This is seen by Clifford algebraists as a linear combination of a scalar, u_0 , a vector, $u_1\vec{e_1}+u_2\vec{e_2}$, and a bi-vector $u_{12}\vec{e_{12}}$.[31] Combining two basis vectors together as $u_1\vec{e_1}+u_2\vec{e_2}$ is done because it fits the interpretation that the basis vectors are the basis of a $\mathbb{R}^n$ type of space. We will challenge this interpretation later, and so we will not follow this particular Clifford algebra convention.

It is not hard to imagine a Clifford algebra with three or more basis vectors all forming anti-commutative bi-vectors, tri-vectors etc.. For any number of basis vectors, once we have the squares (plus unity or minus unity) of those basis vectors and the relation that any product of different basis vectors is anti-commutative, we have the entire algebra. That's worth repeating.

For emphasis:
For any number of basis vectors, once we have the squares (plus unity or minus unity) of those basis vectors and the relation that any product of different basis vectors is anti-commutative, we have the entire algebra.

Hang on! Once we have a set of square roots of plus or minus unity, we have a division algebra basis. You see the connection.

Sub-algebras of the Clifford algebras:
The identity, the real scalar, 1, is the basis of a 1-dimensional sub-algebra, the real numbers, of any Clifford algebra. Together with the identity, 1, each element of the Clifford algebra $Cl_{2,0}$ forms the basis of a 2-dimensional sub-algebra. We have:

$$\{1,\vec{e_1}\} \quad : \quad 1\times1=1, \quad 1\times\vec{e_1}=\vec{e_1}\times1=\vec{e_1}, \quad \vec{e_1}\times\vec{e_1}=1 \qquad (3.12)$$

---

[31] The bi-vector is sometimes referred to as a pseudoscalar. It is clearly not a vector, and it is the square root of minus unity, and so it is a number, but it is not a 'normal' real number, and so it is a 'pseudo' number.

With similar multiplicative closure for $\{1, \vec{e_2}\}$ and $\{1, \vec{e_{12}}\}$. In most Clifford algebra texts, these sub-algebras are not used and are hardly mentioned because they do not correspond with well-known division algebras. These sub-algebras correspond to the hyperbolic complex numbers discovered by James Cockle in 1948, and these are a division algebra in only their polar form. That is why they are traditionally ignored.

The 8-dimensional Clifford algebras, for example, $Cl_{3,0}$, have seven 4-dimensional sub-algebras as well as the seven 2-dimensional sub-algebras, and the 16-dimensional Clifford algebras like $Cl_{4,0}$ have fifteen 2-dimensional sub-algebras, thirty-five 4-dimensional sub-algebras, and fifteen 8-dimensional sub-algebras etc..

Chapter 4

The Traditional Derivation of the Clifford Algebras

Although the geometric concepts of vectors and space are not of the algebraic essence of Clifford algebras, historically, these geometric concepts were used to derive the Clifford algebras. We follow that path.

We seek a way to multiply vectors together that satisfies the same axioms (has the same properties) as the multiplication of real numbers, distributivity, associativity, and commutativity, and we require that the norm of the vectors is preserved in multiplication:

$$\left| \vec{ab} \right| = \left| \vec{a} \right| \left| \vec{b} \right| \tag{4.1}$$

In $\mathbb{R}^n$, the above requirements of distributivity, associativity, and commutativity are not all possible in more than two dimensions, and so we drop the requirement of commutativity. Looking at preserving the norm, we take the square of a vector and we seek a vector product that has this equal to the norm; we seek:

$$\left(x\vec{e_1} + y\vec{e_2} \right)^2 = x^2 + y^2 \tag{4.2}$$

This, (4.2) is just the Pythagoras theorem. The expression on thr LHS of (4.2) is the hypoteneus of a right-angled triangle.

Using the distributive rule, without assuming commutativity, we obtain:

$$x^2 \vec{e_1} + y^2 \vec{e_2} + xy\left(\vec{e_1}\vec{e_2} + \vec{e_2}\vec{e_1} \right) = x^2 + y^2 \tag{4.3}$$

This is satisfied if the orthogonal unit vectors $\vec{e_1}, \vec{e_2}$ obey the multiplication rules:

$$\vec{e_1}^{\,2} = 1, \qquad \vec{e_2}^{\,2} = 1, \qquad \vec{e_1}\vec{e_2} = -\vec{e_2}\vec{e_1} \qquad (4.4)$$

Use associativity to calculate the square

$$\left(\vec{e_1}\vec{e_2}\right)^2 = \vec{e_1}\vec{e_2}\vec{e_1}\vec{e_2} = -\vec{e_1}\vec{e_2}\vec{e_2}\vec{e_1}$$

$$= -\vec{e_1}\left(\vec{e_2}\vec{e_2}\right)\vec{e_1} = -\vec{e_1}.1.\vec{e_1} = -\left(\vec{e_1}\right)^2 = -1 \qquad (4.5)$$

This algebra is known as $Cl_{2,0}$. Since the square of the product $\vec{e_1}\vec{e_2}$ is negative, it was taken that $\vec{e_1}\vec{e_2}$ is neither a scalar nor a vector. Clifford saw this product as a new kind of unit called a bi-vector. The reader might think this new kind of unit is simply the imaginary square root of minus unity, $i = \sqrt{-1}$,[32] but, since there are two basis vectors in $Cl_{2,0}$, Clifford algebraists take the view that they are working in $\mathbb{R}^2$ and there is a reluctance to introduce the imaginary number $i = \sqrt{-1}$ into $\mathbb{R}^2$. It is usual to refer to the above algebra as the Clifford algebra of the vector plane $\mathbb{R}^2$.

Aside:
Within Clifford algebras, intuitively, bi-vectors are associated with rotation. This is because bi-vectors are often associated with imaginary square roots of minus unity, which is a rotation in the complex plane. The relationship is not clear because bi-vectors are sometimes the square roots of plus unity, which is a rotation in space-time.

Another aside:
We see that non-commutativity was derived by Clifford from the desire to produce a product of two vectors that 'made sense' and maintained the norm (distance function). However, non-commutativity occurs far more naturally within matrices and particularly for us within the algebras that derive from the

---

[32] Your author would agree.

commutative $C_2 \times C_2 \times \dots$ finite groups[33]; we do not need the 'product of vectors approach' to gain non-commutativity.

Back to the Clifford algebras:

We can similarly derive a 8-dimensional Clifford algebra, $Cl_{3,0}$, by insisting that:

$$\left(x\vec{e_1} + y\vec{e_2} + z\vec{e_3}\right)^2 = x^2 + y^2 + z^2$$

$$\left.\begin{cases} x^2\vec{e_1}\vec{e_1} + xy\vec{e_1}\vec{e_2} + xz\vec{e_1}\vec{e_3} + xy\vec{e_2}\vec{e_1} + y^2\vec{e_2}\vec{e_2} \\ +yz\vec{e_2}\vec{e_3} + xz\vec{e_3}\vec{e_1} + yz\vec{e_3}\vec{e_2} + z^2\vec{e_3}\vec{e_3} \end{cases}\right\} = x^2 + y^2 + z^2 \qquad (4.6)$$

$$\vec{e_i}\vec{e_i} = 1$$

$$\vec{e_i}\vec{e_j} = -\vec{e_j}\vec{e_i} \quad : \quad j \neq k$$

In Clifford's eyes, we are just using the Pythagoras theorem in higher dimensional spaces.

Or, we can derive a different 8-dimensional Clifford algebra, $Cl_{2,1}$ from insisting that:

$$\left(x\vec{e_1} + y\vec{e_2} + z\vec{e_3}\right)^2 = x^2 + y^2 - z^2 \qquad (4.7)$$

In this case, $\left(\vec{e_3}\right)^2 = -1$ [34]. All of the conventional Clifford algebras can be derived by analogous means. We are extending the Pythagoras theorem into space-time.

---

[33] Although the $C_2 \times C_2 \times \dots$ finite groups are commutative, they underlie non-commutative division algebras such as the quaternions.

[34] We have a vector that is a square root of minus unity; this rather blurs the distinction between vectors and bi-vectors or other multi-vectors.

Aside:

The whole issue of generating Clifford algebras by assuming conservation of the norm is questionable in the 3-dimensional form (and elsewhere) because, as proved by Lagrange in circa 1830[35]:

$$\left(x^2 + y^2 + z^2\right)\left(a^2 + b^2 + c^2\right) \neq R^2 + S^2 + T^2 \qquad (4.8)$$

The Clifford product:

The origin of the Clifford product is simple multiplication of vectors given the nature of the relations like (4.4) above. For $Cl_{2,0}$, the Clifford product, $\overline{\overline{ab}}$, of two vectors $\{\vec{a}, \vec{b}\}$ is (we must maintain the order of the vectors):

$$\overline{\overline{ab}} = \left(a_1\vec{e_1} + a_2\vec{e_2}\right)\left(b_1\vec{e_1} + b_2\vec{e_2}\right)$$
$$= a_1b_1\vec{e_1}\vec{e_1} + a_2b_2\vec{e_2}\vec{e_2} + \left(a_1b_2\vec{e_1}\vec{e_2} + a_2b_1\vec{e_2}\vec{e_1}\right) \qquad (4.9)$$
$$= a_1b_1 + a_2b_2 + \left(a_1b_2 - a_2b_1\right)\vec{e_{12}}$$

We have chosen to denote the Clifford product as $\overline{\overline{ab}}$; this is unconventional. For $Cl_{1,1}$, the Clifford product reflects the fact that one of the basis vectors is a square root of minus unity (a sign difference):

$$\overline{\overline{ab}} = \left(a_1\vec{e_1} + a_2\vec{e_2}\right)\left(b_1\vec{e_1} + b_2\vec{e_2}\right)$$
$$= a_1b_1\vec{e_1}\vec{e_1} + a_2b_2\vec{e_2}\vec{e_2} + \left(a_1b_2\vec{e_1}\vec{e_2} + a_2b_1\vec{e_2}\vec{e_1}\right) \qquad (4.10)$$
$$= a_1b_1 - a_2b_2 + \left(a_1b_2 - a_2b_1\right)\vec{e_{12}}$$

---

[35] There are no 3×3 matrices that are of multiplicative closed form whose determinant is a Euclidean quadratic form. The only 3×3 multiplicatively closed matrices are based on the C_3 finite group with determinant of the form $a^3 + b^3 + c^3 - 3abc$.

The Clifford product is the sum of a scalar and a bi-vector[36]. This is the vector product which Clifford sought. Note: $\overrightarrow{e_{12}}$ is an abbreviation for the bi-vector $\overrightarrow{e_1 e_2}$. The Clifford product is written as:

$$\overline{\overline{ab}} = \vec{a} \cdot \vec{b} + \vec{a} \wedge \vec{b} \qquad (4.11)$$

We can improperly[37] think of the wedge product as the cross-product of two vectors, and so we have both the dot-product and the cross-product within the Clifford product. We also have:

$$\overline{\overline{ba}} = \vec{a} \cdot \vec{b} - \vec{a} \wedge \vec{b} \qquad (4.12)$$

We now have a definition of the product of two vectors. The product is the simple distributive and associative product with which we are all familiar with the appropriate relations analogous to (4.4). The product is a linear product – it is the kind of product we get as elements of a matrix product.

We see that the dot product and the wedge product of the two vectors is given in terms of the Clifford product as:

$$\vec{a} \cdot \vec{b} = \frac{1}{2}\left(\overline{\overline{ab}} + \overline{\overline{ba}}\right)$$
$$\vec{a} \wedge \vec{b} = \frac{1}{2}\left(\overline{\overline{ab}} - \overline{\overline{ba}}\right) \qquad (4.13)$$

Aside:
We have the product of a conjugate complex number with another complex number, the inner product, as:

---

[36] Of course, the inner product in 2-dimensional space-time, $a_1 b_1 - a_2 b_2$, is a scalar.

[37] The cross product of two vectors is a vector in only 3-dimensional space and in 7-dimensional space. The wedge product is a different object. That is why identifying the two objects is improper.

$$\begin{bmatrix} a_1 & -a_2 \\ a_2 & a_1 \end{bmatrix}\begin{bmatrix} b_1 & b_2 \\ -b_2 & b_1 \end{bmatrix} = \begin{bmatrix} a_1b_1 + a_2b_2 & a_1b_2 - a_2b_1 \\ -(a_1b_2 - a_2b) & a_1b_1 + a_2b_2 \end{bmatrix} \tag{4.14}$$

Which the reader can compare to the Clifford product, (4.9).

Elements of the $Cl_{3,0}$ algebra like $\overrightarrow{e_{12}}$, $\overrightarrow{e_{13}}$, $\overrightarrow{e_{23}}$ are basis bi-vectors, and so, conventionally, analogously to vectors, Clifford algebraists might form a general bi-vector[38] as a linear sum of these elements:

$$\vec{B} = b_1\overrightarrow{e_{12}} + b_2\overrightarrow{e_{13}} + b_3\overrightarrow{e_{23}} \tag{4.15}$$

We can form the Clifford product of a general vector, $\bar{a}$, with a general bi-vector, $\vec{B}$. We demonstrate with the $Cl_{3,0}$ algebra:

$$\overline{\overline{aB}} = \vec{a}\vec{B} = \left(a_1\vec{e_1} + a_2\vec{e_2} + a_3\vec{e_3}\right)\left(b_1\overrightarrow{e_{12}} + b_2\overrightarrow{e_{13}} + b_3\overrightarrow{e_{23}}\right)$$

$$= \ldots + a_1 b_3 \overrightarrow{e_1 e_{23}} + \ldots = \ldots + a_1 b_1 \overrightarrow{e_{123}} + \ldots \tag{4.16}$$

The wedge product of a vector with a bi-vector is then given by:

$$\vec{a} \wedge \vec{B} = \frac{1}{2}\left(\overline{\overline{aB}} + \overline{\overline{Ba}}\right) \tag{4.17}$$

This is a 3-vector (the vector bits cancel).

The contraction of a vector with a bi-vector is then given by:

$$\vec{a}\rfloor\vec{B} = \frac{1}{2}\left(\overline{\overline{aB}} - \overline{\overline{Ba}}\right) = -\vec{B}\lfloor\vec{a} \tag{4.18}$$

This is a vector (the 3-vector bits cancel).

If we take the view that there is no algebraic difference between a vector and a bi-vector, then contracting a vector with a bi-vector to form a vector is like crossing two vectors together to form a vector. Contraction is of no great interest to us in this book.

---

[38] A general vector is a linear sum of the basis vectors. A general bi-vector is a linear sum of the basis bi-vectors.

Chapter 5

More Traditional Clifford Algebra

A 4-dimensional algebra from two basis elements:

In the case of $Cl_{2,0}$, we now have a 4-dimensional algebra, but it is a 4-dimensional algebra that has arisen from only two 2-dimensional vectors and is associated with the distance function $d^2 = x^2 + y^2$. There is something that feels wrong to your author about having a 4-dimensional algebra associated with a 2-dimensional space. Within the division algebras, we always have a n-dimensional space associated with a n-dimensional algebra. What is really happening is that the entire $Cl_{2,0}$ algebra is determined by the relations (4.4) between only two elements of that algebra which we have identified with the two basis vectors. The two basis vectors generate the entire algebra. This is not unique to Clifford algebras. We can generate the complex numbers, $\mathbb{C}$, with no more than the single basis element $i = \sqrt{-1}$. We will shortly meet a 4-dimensional A_3 division algebra, *SSA*, which is similarly determined by only two basis elements and in which the relations (4.4) appear between these basis elements within the algebraic matrix form. This concept of generating basis elements is very closely related to the same concept within a finite group. A finite group of order N is generated by less than N individual elements of the finite group.

Matrix representations:

It is established mathematics that the Clifford algebras can be written as matrices. We will not be much concerned with this aspect of the Clifford algebras in this work because, although we use matrices, we use them in a way that is very different to the way they are usually used in Clifford algebra. However, we will simply state a couple of isomorphisms. The Clifford algebra $Cl_{2,0}$ is isomorphic to the algebra of 2×2 matrices with real elements, $Mat(2,\mathbb{R})$. This means we have the equivalences:

$$Cl_{2,0} \cong Mat(2,\mathbb{R})$$

$$1 \equiv \begin{bmatrix} 1 & 0 \\ 0 & 1 \end{bmatrix}, \qquad \overrightarrow{e}_1 \equiv \begin{bmatrix} 1 & 0 \\ 0 & -1 \end{bmatrix}$$

$$\overrightarrow{e}_2 \equiv \begin{bmatrix} 0 & 1 \\ 1 & 0 \end{bmatrix}, \qquad \overrightarrow{e}_{12} \equiv \begin{bmatrix} 0 & 1 \\ -1 & 0 \end{bmatrix}$$

(5.1)

The Clifford algebra $Cl_{3,0}$ is isomorphic to an algebra of 2×2 matrices with complex elements, $Mat(2,\mathbb{C})$. The astute reader might realise that, since the Pauli matrices, σ_i, generate $Mat(2,\mathbb{C})$, $Cl_{3,0}$ can be represented by the Pauli matrices and the products of the Pauli matrices; we have the equivalences:

$$
\begin{array}{cc}
I & 1 \\
\sigma_1, \sigma_2, \sigma_3 & \overrightarrow{e}_1, \overrightarrow{e}_2, \overrightarrow{e}_3 \\
\sigma_1\sigma_2, \sigma_1\sigma_3, \sigma_2\sigma_3 & \overrightarrow{e}_{12}, \overrightarrow{e}_{13}, \overrightarrow{e}_{23} \\
\sigma_1\sigma_2\sigma_3 & \overrightarrow{e}_{123}
\end{array}
$$

(5.2)

We will cover this in more detail later. Similarly, the four Dirac gamma matrices are of the nature:

$$\gamma_0 = \sqrt{+1}, \; \gamma_1 = \sqrt{-1}, \; \gamma_2 = \sqrt{-1}, \; \gamma_2 = \sqrt{-1}$$

$$\gamma_\mu \gamma_\nu = -\gamma_\nu \gamma_\mu \quad : \quad \mu \neq \nu$$

(5.3)

and so they generate the Clifford algebra $Cl_{1,3}$. The reader might wonder what is so special about $Cl_{1,3}$ that it features so prominently in particle physics; there is nothing special about $Cl_{1,3}$.

Multiplicative inverses in Clifford algebra:

The conventional conjugate of an element of the Clifford algebra, $Cl_{2,0}$:

$$u = u_0 + u_1 \vec{e_1} + u_2 \vec{e_2} + u_{12} \overrightarrow{e_{12}} \qquad (5.4)$$

is:

$$\bar{u} = u_0 - u_1 \vec{e_1} - u_2 \vec{e_2} - u_{12} \overrightarrow{e_{12}} \qquad (5.5)$$

This conjugate is the equivalent in the corresponding matrix algebra to the adjoint matrix.

The inverse of an element of the Clifford algebra is given by:

$$u^{-1} = \frac{\bar{u}}{\overline{u}u} \quad : \quad \overline{u}u \neq 0 \qquad (5.6)$$

Note that $\overline{u}u = \overline{u}u$, and so the multiplicative inverse is unambiguous. It is not always the case that the multiplicative inverse exists because it is not always the case that $\overline{u}u \neq 0$. This is one of the reasons why Clifford algebras do not qualify as division algebras.

The centre of $Cl_{3,0}$:

Not everything in a Clifford algebra is non-commutative. The set of basis elements that commute with every other element is called the centre of the algebra. In $Cl_{3,0}$, the tri-vector $\overrightarrow{e_{123}}$ commutes with every other element. Together with the real scalar, 1, this element forms the centre of $Cl_{3,0}$. We have:

$$Cen\left(Cl_{3,0}\right) = \left\{1, \ \overrightarrow{e_{123}}\right\} \qquad (5.7)$$

Together, these two elements are isomorphic to the complex numbers; we have:

$$Cen\left(Cl_{3,0}\right) \simeq \mathbb{C} \qquad (5.8)$$

In general, within a 8-dimensional Clifford algebra of the form $Cl_{p,q} : p+q = 3 : \{p,q\} \geq 0$, the element $\overrightarrow{e_{123}}$ is multiplicatively commutative. We will see the same phenomenon in the 8-dimensional $C_2 \times C_2 \times C_2$ division algebras.

Even and odd algebras:
Looking at an element of $Cl_{2,0}$:

$$u = u_0 + u_1 \overrightarrow{e_1} + u_2 \overrightarrow{e_2} + u_{12} \overrightarrow{e_{12}} \tag{5.9}$$

we notice that, if $u_1 = u_2 = 0$, then we have the complex numbers, $\mathbb{C}$, and, if $u_0 = u_{12} = 0$, then we have $\mathbb{R}^2$. The part $\{u_0, u_{12}\}$ of the Clifford algebra $Cl_{2,0}$ is called the even sub-algebra, $Cl_{2,0}^+$. We use a superscript plus sign to indicate the even sub-algebra. The part $\{u_1, u_2\}$ of the Clifford algebra $Cl_{2,0}$ is called the odd sub-algebra, $Cl_{2,0}^-$. We use a superscript minus sign to indicate the odd sub-algebra.

The even part of a Clifford algebra are the basis elements that are the products of an even number of vectors. The odd part of a Clifford algebra are the basis elements that are the products of an odd number of vectors. The even elements of a Clifford algebra often form a sub-algebra, but the odd elements never form a sub-algebra because they do not include the identity, 1, and so the odd sub-algebra is badly named.

Within $Cl_{3,0}$, the even sub-algebra is isomorphic to the quaternions; we have:

$$Cl_{3,0}^+ \simeq \mathbb{H} \tag{5.10}$$

The reader might think that it is indeed a peculiar thing that we have both $\mathbb{R}^3$ and $\mathbb{H}$ in the same algebra; are we doing complex numbers or are we doing $\mathbb{R}^3$ algebra? This dichotomy is what leads your author to rewrite Clifford algebra.

Spinors in conventional Clifford algebra:

Conventionally, the even sub-algebra of a Clifford algebra is seen as the set of spinors of that Clifford algebra. The even sub-algebra of a Clifford algebra contains the spin group of that Clifford algebra which is the set of unit length elements of the even sub-algebra. These unit length elements of the even sub-algebra are seen as normalised spinors. As we have seen above, the quaternions are a double cover of the 'normal' 2-dimensional rotations in $\mathbb{R}^n$. This is what motivates the identification of the even sub-algebras with spinor algebras.

Since, within Clifford algebra, the even sub-algebras are often identified with spinor algebras, and the complex numbers, $\mathbb{C}$, are an even sub-algebra but not a double cover, there is ambiguity about whether or not a complex number is a spinor.

We will see that this obsession with the even sub-algebra is misplaced and that, when written as a division algebra, the whole Clifford algebra is a set of spinors and the spin group is a n-dimensional rotation matrix.

Different distance functions in $\mathbb{R}^n$:

We see above (4.2) that we derived the Clifford algebra $Cl_{2,0}$ by putting the square of a 2-dimensional vector equal to the 2-dimensional quadratic distance function. We can similarly derive other Clifford algebras by equating the square of a n-dimensional vector to other quadratic n-dimensional distance functions. The signature of the distance function does not have to be all pluses, and we get Clifford algebra like $Cl_{1,3}$ from distance functions with signatures that are not all pluses. However, the conventional Clifford algebras are constrained to use a quadratic form for the distance function.

There is a thing of great interest here. We will see that the 8-dimensional Clifford algebras are really just the 8-dimensional $C_2 \times C_2 \times C_2$ division algebras. These 8-dimensional division algebras do not have a quadratic form as their distance function. It is remarkable that these 8-dimensional algebras are generated by basis elements that do have a quadratic form for their distance function. This is not clearly understood, but it seems to be connected to the existence of 4-dimensional sub-algebras. We discuss this later in this text.

Another 4-dimensional Clifford algebra:

The Clifford algebra, $Cl_{1,1}$, of the 2-dimensional distance function with signature $(+,-)$ is:

$$\left(x\vec{e_1} + y\vec{e_2} \right)^2 = x^2 - y^2 \tag{5.11}$$

$$x^2\vec{e_1} + y^2\vec{e_2} + xy\left(\vec{e_1}\vec{e_2} + \vec{e_2}\vec{e_1} \right) = x^2 - y^2 \tag{5.12}$$

This is satisfied if the orthogonal unit vectors $\vec{e_1},\vec{e_2}$ obey the multiplication rules

$$\vec{e_1}^2 = 1, \ \vec{e_2}^2 = -1, \ \vec{e_1}\vec{e_2} = -\vec{e_2}\vec{e_1} \tag{5.13}$$

Calculate the square:

$$\left(\vec{e_1}\vec{e_2} \right)^2 = \vec{e_1}\vec{e_2}\vec{e_1}\vec{e_2} = -\vec{e_1}\vec{e_1}\vec{e_2}\vec{e_2}$$
$$= -\left(\vec{e_1} \right)^2 \left(\vec{e_2} \right)^2 = +1 \tag{5.14}$$

Having obtained the real scalar, we see that we have a 4-dimensional Clifford algebra with two imaginary square roots of plus unity and one imaginary square root of minus unity.

$$a + b\vec{e_1} + c\vec{e_2} + d\vec{e_1}\vec{e_2}$$
$$a + b\sqrt{+1} + c\sqrt{-1} + d\sqrt{+1} \tag{5.15}$$

Higher dimensional Clifford algebras:

With only a shallow look, we might think that there are Clifford algebras with any number of square roots of minus unity and any number of square roots of plus unity; this is not so; the nature of all

38

the bi-vectors and other multi-vectors is determined by the nature of the generating basis vectors and that determination does not allow any random assortment of square roots of minus unity and square roots of plus unity.

The distance functions used to generate the Clifford algebras have to be quadratic forms, and there are a limited number of possible signatures of these quadratic forms. There are two 1-dimensional quadratic distance functions, $\{d^2 = x^2, d^2 = -x^2\}$. Putting the square of the single basis vector equal to d^2 leads to two 2-dimensional algebras:

$$Cl_{1,0} \simeq \{1, \ \vec{e_1} = \sqrt{+1}\} \quad \& \quad Cl_{0,1} \simeq \{1, \ \vec{e_1} = \sqrt{-1}\} \qquad (5.16)$$

Yes! $Cl_{0,1}$ is algebraically isomorphic to $\mathbb{C}$. $Cl_{1,0}$ is almost algebraically isomorphic to the hyperbolic complex numbers, $\mathbb{S}$, but since we have to take the exponential of the basic matrix form to get the hyperbolic complex numbers, strictly speaking we cannot say $Cl_{1,0}$ is algebraically isomorphic to $\mathbb{S}$. Both of these algebras are commutative, and so, strictly speaking, these are not Clifford algebras. We include them for the sake of completeness and because they help to show the general pattern of the Clifford algebras.

There are three distinct 2-dimensional quadratic distance functions corresponding to three distinct 4-dimensional Clifford algebras:

$$Cl_{2,0} \simeq d^2 = x^2 + y^2 \quad : \quad 1, \ \vec{e_1} = \sqrt{+1}, \ \vec{e_2} = \sqrt{+1}, \ \vec{e_{12}} = \sqrt{-1}$$
$$Cl_{0,2} \simeq d^2 = -x^2 - y^2 \quad : \quad 1, \ \vec{e_1} = \sqrt{-1}, \ \vec{e_2} = \sqrt{-1}, \ \vec{e_{12}} = \sqrt{-1}$$
$$Cl_{1,1} \simeq d^2 = x^2 - y^2 \quad : \quad 1, \ \vec{e_1} = \sqrt{+1}, \ \vec{e_2} = \sqrt{-1}, \ \vec{e_{12}} = \sqrt{+1}$$
$$(5.17)$$

Yes! $Cl_{0,2}$ is algebraically isomorphic to the quaternions, $\mathbb{H}$. We have an algebra with three square roots of minus unity, and two algebras with two square roots of plus unity and one square root of minus unity. The reader who is wise in the lore of the $C_2 \times C_2 \times ...$ algebras will realise these combinations correspond to the 4-dimensional non-commutative division algebras. In a table:

$$\begin{array}{cc} \sqrt{+1} & \sqrt{-1} \\ 0 & 3 \\ 2 & 1 \end{array} \tag{5.18}$$

There are four distinct 3-dimensional quadratic distance functions corresponding to four 8-dimensional Clifford algebras[39]:

$$Cl_{3,0} \simeq d^2 = x^2 + y^2 + z^2 \quad : \quad 1, \ \vec{e_1} = \sqrt{+1}, \ \vec{e_2} = \sqrt{+1}, \ \vec{e_3} = \sqrt{+1},$$
$$\vec{e_{12}} = \sqrt{-1}, \ \vec{e_{13}} = \sqrt{-1}, \ \vec{e_{23}} = \sqrt{-1}, \ \vec{e_{123}} = \sqrt{-1} \tag{5.19}$$

$$Cl_{0,3} \simeq d^2 = -x^2 - y^2 - z^2 \quad : \quad 1, \ \vec{e_1} = \sqrt{-1}, \ \vec{e_2} = \sqrt{-1}, \ \vec{e_3} = \sqrt{-1},$$
$$\vec{e_{12}} = \sqrt{-1}, \ \vec{e_{13}} = \sqrt{-1}, \ \vec{e_{23}} = \sqrt{-1}, \ \vec{e_{123}} = \sqrt{+1} \tag{5.20}$$

$$Cl_{2,1} \simeq d^2 = x^2 + y^2 - z^2 \quad : \quad 1, \ \vec{e_1} = \sqrt{+1}, \ \vec{e_2} = \sqrt{+1}, \ \vec{e_3} = \sqrt{-1},$$
$$\vec{e_{12}} = \sqrt{-1}, \ \vec{e_{13}} = \sqrt{+1}, \ \vec{e_{23}} = \sqrt{+1}, \ \vec{e_{123}} = \sqrt{+1} \tag{5.21}$$

$$Cl_{1,2} \simeq d^2 = x^2 - y^2 - z^2 \quad : \quad 1, \ \vec{e_1} = \sqrt{+1}, \ \vec{e_2} = \sqrt{-1}, \ \vec{e_3} = \sqrt{-1},$$
$$\vec{e_{12}} = \sqrt{+1}, \ \vec{e_{13}} = \sqrt{+1}, \ \vec{e_{23}} = \sqrt{-1}, \ \vec{e_{123}} = \sqrt{-1} \tag{5.22}$$

In a table, these are:

---

[39] There is a technical complication with algebras in which the number of plus signs, p, less the number of minus signs, q, in the distance function is such that $p - q = 1 \mod(4)$. Such algebras are semi-simple Clifford algebras rather than simple Clifford algebras, but this does not concern us here.

$$\sqrt{+1} \quad \sqrt{-1}$$

$$
\begin{array}{cc}
3 & 4 \\
1 & 6 \\
5 & 2
\end{array}
$$

$$(5.23)$$

There are five 4-dimensional quadratic distance functions corresponding to five 16-dimensional Clifford algebras:

$$Cl_{4,0} \simeq d^2 = t^2 + x^2 + y^2 + z^2$$

$$1, \quad \vec{e}_1 = \sqrt{+1}, \quad \vec{e}_2 = \sqrt{+1}, \quad \vec{e}_3 = \sqrt{+1}, \quad \vec{e}_4 = \sqrt{+1},$$

$$\vec{e}_{12} = \sqrt{-1}, \quad \vec{e}_{13} = \sqrt{-1}, \quad \vec{e}_{14} = \sqrt{-1}, \quad \vec{e}_{23} = \sqrt{-1}, \quad \vec{e}_{24} = \sqrt{-1}, \quad \vec{e}_{34} = \sqrt{-1},$$

$$\vec{e}_{123} = \sqrt{-1}, \quad \vec{e}_{124} = \sqrt{-1}, \quad \vec{e}_{134} = \sqrt{-1}, \quad \vec{e}_{234} = \sqrt{-1}, \quad \vec{e}_{1234} = \sqrt{+1}$$

$$(5.24)$$

$$Cl_{0,4} \simeq d^2 = -t^2 - x^2 - y^2 - z^2$$

$$1, \quad \vec{e}_1 = \sqrt{-1}, \quad \vec{e}_2 = \sqrt{-1}, \quad \vec{e}_3 = \sqrt{-1}, \quad \vec{e}_4 = \sqrt{-1},$$

$$\vec{e}_{12} = \sqrt{-1}, \quad \vec{e}_{13} = \sqrt{-1}, \quad \vec{e}_{14} = \sqrt{-1}, \quad \vec{e}_{23} = \sqrt{-1}, \quad \vec{e}_{24} = \sqrt{-1}, \quad \vec{e}_{34} = \sqrt{-1},$$

$$\vec{e}_{123} = \sqrt{+1}, \quad \vec{e}_{124} = \sqrt{+1}, \quad \vec{e}_{134} = \sqrt{+1}, \quad \vec{e}_{234} = \sqrt{+1}, \quad \vec{e}_{1234} = \sqrt{+1}$$

$$(5.25)$$

$$Cl_{3,1} \simeq d^2 = t^2 + x^2 + y^2 - z^2$$

$$1, \quad \vec{e}_1 = \sqrt{+1}, \quad \vec{e}_2 = \sqrt{+1}, \quad \vec{e}_3 = \sqrt{+1}, \quad \vec{e}_4 = \sqrt{-1},$$

$$\vec{e}_{12} = \sqrt{-1}, \quad \vec{e}_{13} = \sqrt{-1}, \quad \vec{e}_{14} = \sqrt{+1}, \quad \vec{e}_{23} = \sqrt{-1}, \quad \vec{e}_{24} = \sqrt{+1}, \quad \vec{e}_{34} = \sqrt{+1},$$

$$\vec{e}_{123} = \sqrt{-1}, \quad \vec{e}_{124} = \sqrt{+1}, \quad \vec{e}_{134} = \sqrt{+1}, \quad \vec{e}_{234} = \sqrt{+1}, \quad \vec{e}_{1234} = \sqrt{-1}$$

$$(5.26)$$

$$Cl_{2,2} \simeq d^2 = t^2 + x^2 - y^2 - z^2$$

$$1, \quad \vec{e}_1 = \sqrt{+1}, \quad \vec{e}_2 = \sqrt{+1}, \quad \vec{e}_3 = \sqrt{-1}, \quad \vec{e}_4 = \sqrt{-1},$$

$$\vec{e}_{12} = \sqrt{-1}, \quad \vec{e}_{13} = \sqrt{+1}, \quad \vec{e}_{14} = \sqrt{+1}, \quad \vec{e}_{23} = \sqrt{+1}, \quad \vec{e}_{24} = \sqrt{+1}, \quad \vec{e}_{34} = \sqrt{-1},$$

$$\vec{e}_{123} = \sqrt{+1}, \quad \vec{e}_{124} = \sqrt{+1}, \quad \vec{e}_{134} = \sqrt{-1}, \quad \vec{e}_{234} = \sqrt{-1}, \quad \vec{e}_{1234} = \sqrt{+1}$$

$$(5.27)$$

$$Cl_{1,3} \cong d^2 = t^2 - x^2 - y^2 - z^2$$

$$1, \ \vec{e_1} = \sqrt{+1}, \ \vec{e_2} = \sqrt{-1}, \ \vec{e_3} = \sqrt{-1}, \ \vec{e_4} = \sqrt{-1},$$

$$\vec{e_{12}} = \sqrt{+1}, \ \vec{e_{13}} = \sqrt{+1}, \ \vec{e_{14}} = \sqrt{+1}, \ \vec{e_{23}} = \sqrt{-1}, \ \vec{e_{24}} = \sqrt{-1}, \ \vec{e_{34}} = \sqrt{-1},$$

$$\vec{e_{123}} = \sqrt{-1}, \ \vec{e_{124}} = \sqrt{-1}, \ \vec{e_{134}} = \sqrt{-1}, \ \vec{e_{234}} = \sqrt{+1}, \ \vec{e_{1234}} = \sqrt{-1}$$

$$(5.28)$$

We have 16-dimensional Clifford algebras with the following numbers of imaginary square roots:

$$
\begin{array}{cc}
\sqrt{+1} & \sqrt{-1} \\
5 & 10 \\
9 & 6
\end{array}
\qquad (5.29)
$$

We thus have:

2 dimensions (2 potential algebras)	$Cl_{1,0} \sim \{1, \sqrt{+1}\}$
	$Cl_{0,1} \sim \{1, \sqrt{-1}\}$
4 dimensions (2 potential algebras)	$Cl_{2,0} \sim \{1, 2\sqrt{+1}, \sqrt{-1}\}$
	$Cl_{1,1} \sim \{1, 2\sqrt{+1}, \sqrt{-1}\}$
	$Cl_{0,2} \sim \{1, 3\sqrt{-1}\}$
8-dimensions (3 potential algebras)	$Cl_{3,0} \sim \{1, 3\sqrt{+1}, 4\sqrt{-1}\}$
	$Cl_{2,1} \sim \{1, 5\sqrt{+1}, 2\sqrt{-1}\}$
	$Cl_{1,2} \sim \{1, 3\sqrt{+1}, 4\sqrt{-1}\}$
	$Cl_{0,3} \sim \{1, \sqrt{+1}, 6\sqrt{-1}\}$

16 dimensions (2 potential algebras)	$Cl_{4,0} \sim \{1, 5\sqrt{+1}, 10\sqrt{-1}\}$
	$Cl_{3,1} \sim \{1, 9\sqrt{+1}, 6\sqrt{-1}\}$
	$Cl_{2,2} \sim \{1, 9\sqrt{+1}, 6\sqrt{-1}\}$
	$Cl_{1,3} \sim \{1, 5\sqrt{+1}, 10\sqrt{-1}\}$
	$Cl_{0,4} \sim \{1, 5\sqrt{+1}, 10\sqrt{-1}\}$

Aside:

With one eye on the 16-dimensional Clifford algebras above, we give a standard Clifford algebra theorem[40]:

$$Cl_{p,q} \simeq Cl_{q+1,p-1}$$
$$Cl_{p,q} \simeq Cl_{p-4,q+4}$$

(5.30)

In general:

In general, with thought, we see that the numbers of vectors, bi-vectors, tri-vectors ... follows a pattern. We have:

	Scalars	Vectors	Bi-vectors	3-vectors	4-vectors	5-vectors
2 Dim	1	1				
4 Dim	1	2	1			
8 Dim	1	3	3	1		
16 Dim	1	4	6	4	1	
32 Dim	1	5	10	10	5	1

---

[40] Lounesto page 215 & 216.

Further, the squares (plus or minus unity) of these elements will be determined by the squares of the vectors. The squares of bi-vectors are of the form:

$$\overrightarrow{e_a}\overrightarrow{e_b}\overrightarrow{e_a}\overrightarrow{e_b} = \overrightarrow{e_{ab}}\overrightarrow{e_{ab}} = -\overrightarrow{e_{aa}}\overrightarrow{e_{bb}} = -\left(\overrightarrow{e_a}\right)^2\left(\overrightarrow{e_b}\right)^2 \qquad (5.31)$$

Similarly:

$$\overrightarrow{e_{abc}}\overrightarrow{e_{abc}} = +\overrightarrow{e_{aa}}\overrightarrow{e_{bc}}\overrightarrow{e_{bc}} = -\overrightarrow{e_{aa}}\overrightarrow{e_{bb}}\overrightarrow{e_{cc}} = -\left(\overrightarrow{e_a}\right)^2\left(\overrightarrow{e_b}\right)^2\left(\overrightarrow{e_c}\right)^2$$

$$\overrightarrow{e_{abcd}}\overrightarrow{e_{abcd}} = -\overrightarrow{e_{aa}}\overrightarrow{e_{bcd}}\overrightarrow{e_{bcd}} = -\overrightarrow{e_{aa}}\overrightarrow{e_{bb}}\overrightarrow{e_{cd}}\overrightarrow{e_{cd}}$$

$$= +\overrightarrow{e_{aa}}\overrightarrow{e_{bb}}\overrightarrow{e_{cc}}\overrightarrow{e_{dd}} = +\left(\overrightarrow{e_a}\right)^2\left(\overrightarrow{e_b}\right)^2\left(\overrightarrow{e_c}\right)^2\left(\overrightarrow{e_d}\right)^2 \qquad (5.32)$$

$$\overrightarrow{e_{abcde}}\overrightarrow{e_{abcde}} = +\overrightarrow{e_{aa}}\overrightarrow{e_{bcde}}\overrightarrow{e_{bcde}} = -\overrightarrow{e_{aa}}\overrightarrow{e_{bb}}\overrightarrow{e_{cde}}\overrightarrow{e_{cde}}$$

$$= -\overrightarrow{e_{aa}}\overrightarrow{e_{bb}}\overrightarrow{e_{cc}}\overrightarrow{e_{de}}\overrightarrow{e_{de}} = +\left(\overrightarrow{e_a}\right)^2\left(\overrightarrow{e_b}\right)^2\left(\overrightarrow{e_c}\right)^2\left(\overrightarrow{e_d}\right)^2\left(\overrightarrow{e_e}\right)^2$$

We thus see that the 32-dimensional $Cl_{5,0}$ algebra will have:

1	*Scalar*	$\sqrt{+1}$
5	*vector*	$\sqrt{+1}$
10	*2 – vector*	$\sqrt{-1}$
10	*3 – vector*	$\sqrt{-1}$
5	*4 – vector*	$\sqrt{+1}$
1	*5 – vector*	$\sqrt{+1}$

$$(5.33)$$

That is 20 square roots of minus unity and 12 square roots of plus unity.

The 32 dimensional $Cl_{4,1}$ algebra will have:

1	Scalar	$\sqrt{+1}$		6	3−vector	$\sqrt{-1}$	
4	Vector	$\sqrt{+1}$		4	3−vector	$\sqrt{+1}$	
1	Vector	$\sqrt{-1}$	:	1	4−vector	$\sqrt{+1}$	(5.34)
4	2−vector	$\sqrt{-1}$		4	4−vector	$\sqrt{-1}$	
6	2−vector	$\sqrt{+1}$		1	5−vector	$\sqrt{-1}$	

That is 16 square roots of minus unity and 16 square roots of plus unity.

The 32 dimensional $Cl_{3,2}$ algebra will have:

1	Scalar	$\sqrt{+1}$		4	3−vector	$\sqrt{-1}$	
3	Vector	$\sqrt{+1}$		6	3−vector	$\sqrt{+1}$	
2	Vector	$\sqrt{-1}$	:	3	4−vector	$\sqrt{+1}$	(5.35)
4	2−vector	$\sqrt{-1}$		2	4−vector	$\sqrt{-1}$	
6	2−vector	$\sqrt{+1}$		1	5−vector	$\sqrt{+1}$	

That is 12 square roots of minus unity and 20 square roots of plus unity.

The 32 dimensional $Cl_{2,3}$ algebra will have:

1	Scalar	$\sqrt{+1}$		4	3−vector	$\sqrt{+1}$	
3	Vector	$\sqrt{-1}$		6	3−vector	$\sqrt{-1}$	
2	Vector	$\sqrt{+1}$	:	3	4−vector	$\sqrt{+1}$	(5.36)
4	2−vector	$\sqrt{-1}$		2	4−vector	$\sqrt{-1}$	
6	2−vector	$\sqrt{+1}$		1	5−vector	$\sqrt{-1}$	

That is 16 square roots of minus unity and 16 square roots of plus unity.

The 32 dimensional $Cl_{1,4}$ algebra will have:

1	*Scalar*	$\sqrt{+1}$		6	3 − *vector*	$\sqrt{-1}$
4	*Vector*	$\sqrt{-1}$		4	3 − *vector*	$\sqrt{+1}$
1	*Vector*	$\sqrt{+1}$	:	1	4 − *vector*	$\sqrt{+1}$
4	2 − *vector*	$\sqrt{+1}$		4	4 − *vector*	$\sqrt{-1}$
6	2 − *vector*	$\sqrt{-1}$		1	5 − *vector*	$\sqrt{+1}$

$$(5.37)$$

That is 20 square roots of minus unity and 12 square roots of plus unity.

The 32 dimensional $Cl_{0,5}$ algebra will have:

1	*Scalar*	$\sqrt{+1}$
5	*vector*	$\sqrt{-1}$
10	2 − *vector*	$\sqrt{+1}$
10	3 − *vector*	$\sqrt{-1}$
5	4 − *vector*	$\sqrt{+1}$
1	5 − *vector*	$\sqrt{-1}$

$$(5.38)$$

That is 16 square roots of minus unity and 16 square roots of plus unity.

We have only three distinct 32-dimensional Clifford algebras:

$$Cl_{5,0} \simeq Cl_{1,4}$$
$$Cl_{4,1} \simeq Cl_{2,3} \simeq Cl_{0,5} \qquad (5.39)$$
$$Cl_{3,2}$$

This concurs with (5.30).

Chapter 6

Yet More Traditional Clifford Algebra

Conventional Clifford algebra classification:
All the above, (5.16), (5.18), (5.23), (5.29) & (5.39) concurs with the conventional classification of Clifford algebras, $Cl_{p,q}$, given by Lounesto[41]. The numbers in the left-hand column are $p+q$; the numbers in the top row are $p-q$:

	-4	-3	-2	-1	0	1	2	3	4
0					$\mathbb{R}$				
1				$\mathbb{C}$		$^2\mathbb{R}$			
2			$\mathbb{H}$		$\mathbb{R}(2)$		$\mathbb{R}(2)$		
3		$^2\mathbb{H}$		$\mathbb{C}(2)$		$^2\mathbb{R}(2)$		$\mathbb{C}(2)$	
4	$\mathbb{H}(2)$		$\mathbb{H}(2)$		$\mathbb{R}(4)$		$\mathbb{R}(4)$		$\mathbb{H}(2)$

Sub-algebras again:
In effect, when we count the number of square roots of minus unity and the number of square roots of plus unity, we are counting the relative numbers of the two types of 2-dimensional sub-algebras (5.16) within a particular Clifford algebra. In higher dimensional Clifford algebras, we will see that the relative numbers of the different types of 4-dimensional sub-algebras (or higher dimensional sub-algebras) can also be counted.

---

[41] Lounesto: page 217

Rotation in Clifford algebras:

We can express rotations in either single matrix form where we use only one rotation matrix or in double matrix form where we use two rotation matrices. The rotation group $SO(2)$ is 2-dimensional rotation in $\mathbb{R}^2$. The rotation group $U(1)$ is 2-dimensional rotation in the complex plane $\mathbb{C}$. The two groups are isomorphic as rotation groups. These rotations are traditionally done using the same single 2×2 rotation matrix.

Although, in 2-dimensions, it is entirely sensible to use a single matrix rotation, Clifford algebraists prefer, even in two dimensions, to use double matrix rotations because this will generalise to the non-commutative higher dimensional $\mathbb{R}^n$ spaces[42]. Your author is not sure about this justification, but it is the standard mantra. It is possible to use only one rotation matrix in higher dimensional Clifford algebras, but we then have to accept two different rotations, rotation matrix on the left or rotation matrix on the right, for the same angle.

In short, conventional rotation in Clifford algebras is always a double matrix rotation.

Within the conventional Clifford algebra, $Cl_{2,0}$, we rotate the point $(x,y) \in \mathbb{R}^2$ by rotating the element of the Clifford algebra associated with the basis vectors $(x\vec{e_1} + y\vec{e_2})$. We do this as:

$$\left(\cos\frac{\theta}{2} + \vec{e_{12}}\sin\frac{\theta}{2} \right)^{-1} \left(x\vec{e_1} + y\vec{e_2} \right) \left(\cos\frac{\theta}{2} + \vec{e_{12}}\sin\frac{\theta}{2} \right) \quad (6.1)$$

Note that all the elements in this rotation, (6.1), are elements of the Clifford algebra, but we are mixing $\mathbb{R}^2$ with $\mathbb{C}$. This is (signs):

$$\left(\cos\frac{\theta}{2} - \vec{e_{12}}\sin\frac{\theta}{2} \right) \left(x\vec{e_1} + y\vec{e_2} \right) \left(\cos\frac{\theta}{2} + \vec{e_{12}}\sin\frac{\theta}{2} \right) \quad (6.2)$$

---

[42] Lounesto: Pg 30, footnotes.

Expanding (6.2) gives:

$$\vec{e_1}(x\cos\theta - y\sin\theta) + \vec{e_2}(x\sin\theta + y\cos\theta) \qquad (6.3)$$

This is the same result as:

$$\begin{bmatrix} \cos\theta & \sin\theta \\ -\sin\theta & \cos\theta \end{bmatrix}\begin{bmatrix} x & y \\ -y & x \end{bmatrix} = \begin{bmatrix} x\cos\theta - y\sin\theta & x\sin\theta + y\cos\theta \\ -(x\sin\theta + y\cos\theta) & x\cos\theta - y\sin\theta \end{bmatrix}$$
$$(6.4)$$

In (6.4), we are working entirely within the 2-dimensional complex numbers, $\mathbb{C}$. We have effectively taken the $\mathbb{R}^2$ space of $\left(x\vec{e_1} + y\vec{e_2}\right)$ to be the complex plane and identified $\vec{e_1}$ with the real axis and $\vec{e_2}$ with the imaginary axis.

Because of the even/odd nature of the cosine and sine functions, (6.2) is equal to:

$$\left(\cos\left(-\frac{\theta}{2}\right) + \vec{e_{12}}\sin\left(-\frac{\theta}{2}\right)\right)\left(x\vec{e_1} + y\vec{e_2}\right)\left(\cos\left(-\frac{\theta}{2}\right) - \vec{e_{12}}\sin\left(-\frac{\theta}{2}\right)\right)$$
$$(6.5)$$

Expanding (6.5) gives the same as (6.3):

$$\vec{e_1}(x\cos\theta - y\sin\theta) + \vec{e_2}(x\sin\theta + y\cos\theta) \qquad (6.6)$$

We see that we can rotate through two different angles, $\left\{\frac{\theta}{2}, -\frac{\theta}{2}\right\}$, and have the same rotation. This is called a double cover.

The reader is referred to the 2-dimensional quaternion rotation matrix, (1.3), where, because of the square root, we can get the same rotation by rotating through $\pm b$. The reader should now pause; all the fuss in conventioanal Clifford algebra comes down to a higher dimensional trigonometric function.

Of course, such double matrix multiplication is just the identity in a commutative algebra. This double cover would not exist if we were rotating in only 2-dimensional space, but $Cl_{2,0}$ is a 4-dimensional algebra and we are rotating 2-dimensionally within a 4-dimensional algebra.

Aside:
If we had not used the inverse rotation matrix but had used two copies of the rotation matrix, we would have got only the identity rotation:

$$\left(\cos\frac{\theta}{2} + \overrightarrow{e_{12}}\sin\frac{\theta}{2} \right)\left(x\overrightarrow{e_1} + y\overrightarrow{e_2} \right)\left(\cos\frac{\theta}{2} + \overrightarrow{e_{12}}\sin\frac{\theta}{2} \right)$$

$$= \left(x\overrightarrow{e_1} + y\overrightarrow{e_2} \right)\left(\cos^2\frac{\theta}{2} + \sin^2\frac{\theta}{2} \right) \qquad (6.7)$$

$$= \left(x\overrightarrow{e_1} + y\overrightarrow{e_2} \right)$$

Spin groups conventionally:
Every Clifford algebra has within it a spin group. Conventionally, the spin group of a Clifford algebra is the set of elements of the even-sub-algebra (see: circa (5.9) to (5.10)) of length unity[43]. We can visualise the spin group as the 'spherical surface' that is the set of points unit distance from the origin. Rotation will take any one of these points to another of these points. The spin groups are easily constructed for $\mathbb{R}^{n,p}$ as the set of elements of the Clifford algebra $Cl_{n,p}$ of unit length.[44]

$$spin(n,p) = \left\{ s \in Cl_{n,p} : \tilde{s}s = 1, \overline{s}s = 1 \right\} \qquad (6.8)$$

Examples are[45]:

---

[43] We will modify this when we rewrite the Clifford algebras as division algebras.
[44] Lounesto: page 59.
[45] Lounesto: pages 87, 126, 127, 128

$$spin(4) = \left\{ s \in Cl_{4,0}^{+} : \tilde{s}s = 1 \right\}$$

$$spin(1,3) = \left\{ s \in Cl_{3,0} : \bar{s}s = 1 \right\}$$

$$spin(3,1) = \left\{ s \in Cl_{3,1}^{+} : \tilde{s}s = 1 \right\} \tag{6.9}$$

$$spin(1,3) = \left\{ s \in Cl_{1,3}^{+} : \tilde{s}s = \pm 1 \right\}$$

$$spin_{+}(1,3) = \left\{ s \in Cl_{1,3}^{+} : \tilde{s}s = 1 \right\}$$

These spin groups are 2-fold covers of the special orthogonal groups, $SO(n)$. An element of a spin group is a normalised spinor.

Of course, some Clifford algebras are the even-sub-algebras of higher dimensional Clifford algebras; an example is $Cl_{0,2} \cong \mathbb{H} \cong Cl_{3,0}^{+}$. In such cases, it is conventionally seen that there are two spin groups within the even sub-algebra; one is the unit length elements of the whole even sub-algebra and the other is the unit length elements of the lesser even sub-algebra of the greater even sub-algebra. This is like the complex numbers, $\mathbb{C}$, being a sub-algebra of the quaternions while the quaternions are a sub-algebra of $Cl_{3,0}$. We see already some ambiguity in the definition of a spin group; this will simplified when we rewrite Clifford algebra.

Clifford algebra in a nutshell:
What is effectively happening above is that we choose a number, n, (the dimension of the distance function) that will determine the dimension of the Clifford algebras to be 2^n. We then choose, for each algebra, n objects (basis vectors, operators, whatever we call them) to be either a square root of plus unity or a square root of minus unity and to be anti-commutative with each other.

The particular choices of square roots of plus unity and minus unity of these generating objects then determine the other basis objects (bi-vectors, tri-vectors etc.) to be either a square root of plus unity or a square root of minus unity or a scalar (the product of an object with

51

itself). We see that the whole 2^n-dimensional Clifford algebra is determined by the nature of just n anti-commutative square roots of plus or minus unity – the generators. Within these algebras are spin groups that are the set of unit length elements of the even sub-algebra. The elements of the spin group are normalised spinors. When we rewrite Clifford algebra as the $C_2 \times C_2 \times...$ division algebras, the spin groups will be the rotation matrices of those algebras – much simpler.

Quaternions and $Cl_{0,3}$:

What is a quaternion? It is a set of three multiplicatively connected objects that each square to minus unity and which are all mutually anti-commutative. The 4th power of these objects is unity. What is $Cl_{0,3}$? It is a set of three multiplicatively connected objects that each square to minus unity and which are all mutually anti-commutative. The 4th power of these objects is unity. Who needs basis vectors, bi-vectors etc.?

The Dirac equation etc.:

The Dirac equation has a spinor (bi-spinor) in it. Clearly, if we are to write the Dirac equation as a 'proper' equation within a particular algebra, then that algebra must have spinors within it; it must be a Clifford algebra or a division algebra equivalent to a Clifford algebra.

Summary:

The last three chapters have been quite laborious. In the next chapter, things become much simpler.

Chapter 7

The $C_2 \times C_2 \times ...$ Division Algebras

Division algebras are types of numbers like the real numbers, the complex numbers, or the quaternions. The foundations of division algebras are the real numbers and the finite groups[46]. As an example, we take the regular representation[47] of the finite group C_2 (the 2×2 permutation matrices[48]):

$$\begin{bmatrix} 1 & 0 \\ 0 & 1 \end{bmatrix}, \begin{bmatrix} 0 & 1 \\ 1 & 0 \end{bmatrix} \tag{7.1}$$

We convert the 1's into real variables and add these to get the basic 2-dimensional algebraic matrix form:

$$\begin{bmatrix} a & b \\ b & a \end{bmatrix} \tag{7.2}$$

This matrix satisfies all the axioms of a division algebra (multiplicative closure, multiplicative identity, etc. ...) except that it does not necessarily have a multiplicative inverse because the matrix is not necessarily non-singular. We could take both the positive and negative of the exponential of the basic algebraic matrix form:

$$\pm \exp\left(\begin{bmatrix} a & b \\ b & a \end{bmatrix}\right) = \pm \begin{bmatrix} e^a & 0 \\ 0 & e^a \end{bmatrix} \begin{bmatrix} \cosh b & \sinh b \\ \sinh b & \cosh b \end{bmatrix} = \mathbb{S} \tag{7.3}$$

---

[46] The reader is advised to forget all about algebraic extensions of the real numbers based on unfactorisable monic minimum polynomials – we have it right.
[47] The regular representation of a group of order n is the set of $n \times n$ permutation matrices that have the same multiplicative relations as the elements of the group.
[48] A permutation matrix is a square matrix with a single one in every column and a single one in every row and with zeros everywhere else in the matrix.

This matrix form satisfies all the axioms of a division algebra. (There would be no additive inverse on the real axis if we did not take both the plus and minus forms.) We will not take the negative half; we do not need it, and the maths does not lead to it. We will take only the positive half and not worry about additive inverses on the real axis.

This is the 2-dimensional division algebra most commonly known as the hyperbolic complex numbers – it is 2-dimensional space-time. Because we cannot travel backwards in time, we do not need additive inverses on the real axis. We are then left with the bones of the special theory of relativity which is that physics is invariant under rotation in 2-dimensional space-time (a Lorentz transformation):

$$\begin{bmatrix} \cosh b & \sinh b \\ \sinh b & \cosh b \end{bmatrix} = \begin{bmatrix} \gamma & v\gamma \\ v\gamma & \gamma \end{bmatrix} \qquad (7.4)$$

Scaling parameters:
It is not necessary that the imaginary variable, b , has to scale equally to the real variable (they are independent variables), and so we introduce a scaling parameter and still have a division algebra for all real values of this scaling parameter other than zero:

$$\exp\left(\begin{bmatrix} a & b \\ \lambda b & a \end{bmatrix} \right) \quad : \quad \lambda \neq 0 \qquad (7.5)$$

When the scaling parameter, λ , is positive, we have the hyperbolic complex numbers. When the scaling parameter, λ , is negative, we have the euclidean complex numbers:

$$\exp\left(\begin{bmatrix} a & b \\ -\kappa b & a \end{bmatrix} \right) \quad : \quad \kappa > 0 \equiv \mathbb{C}$$

$$\exp\left(\begin{bmatrix} a & b \\ -b & a \end{bmatrix} \right)_{\kappa=1} = \begin{bmatrix} e^a & 0 \\ 0 & e^a \end{bmatrix} \begin{bmatrix} \cos b & \sin b \\ -\sin b & \cos b \end{bmatrix} = \mathbb{C} \qquad (7.6)$$

The above (7.3) and (7.6) are the only two types of 2-dimensional division algebras[49].

Important aside:

If we ignore the need to exclude singular matrices for the minute, we can write these two algebras as:

$$\mathbb{S} = a + b\sqrt{+1}$$
$$\mathbb{C} = a + b\sqrt{-1}$$
(7.7)

Comparing this to the 2-dimensional Clifford algebras above, (5.16), we see that we have a correspondence between the algebras:

$$Cl_{1,0} \equiv \mathbb{S} \quad : \quad Cl_{0,1} \equiv \mathbb{C}$$
(7.8)

We see that the problem of lack of multiplicative inverse that disqualifies Clifford algebras from being division algebras occurs within the hyperbolic complex numbers but is avoided by taking the polar form.

Higher dimensional complex numbers:

What we have done above with the finite group C_2 can be done with any finite group. The group C_3 will produce four[50] 3-dimensional division algebras based upon cube roots of plus unity and cube roots of minus unity. The group C_4 will produce eight 4-dimensional division algebras based upon fourth roots of plus unity and fourth roots of minus unity.[51]

In this book, we are concerned with only finite groups of the $C_2 \times C_2 \times ...$ form because only these finite groups produce division

---

[49] There is only one scaling parameter because there are only two axes. The scaling parameter can be only positive or negative.

[50] That is four permutations of the plus and minus signs of two scaling parameters.

[51] See Dennis Morris Complex Numbers The Higher Dimensional Forms ISBN: 678-1508877499

algebras which have various numbers of square roots of plus unity and square roots of minus unity and are a match for the Clifford algebras.

Symmetric and anti-symmetric permutation matrices:

The regular representations of the $C_2 \times C_2 \times$... groups are all symmetric permutation matrices. Symmetric permutation matrices always square to plus unity (the identity matrix). By inserting scaling parameters, the symmetric matrices sometimes become anti-symmetric permutation matrices. Anti-symmetric permutation matrices always square to minus unity (minus the identity matrix). For example:

$$
\begin{bmatrix} 0 & 0 & 1 & 0 \\ 0 & 0 & 0 & 1 \\ 1 & 0 & 0 & 0 \\ 0 & 1 & 0 & 0 \end{bmatrix}^2 = \begin{bmatrix} 1 & 0 & 0 & 0 \\ 0 & 1 & 0 & 0 \\ 0 & 0 & 1 & 0 \\ 0 & 0 & 0 & 1 \end{bmatrix}
$$

$$
\begin{bmatrix} 0 & 0 & 1 & 0 \\ 0 & 0 & 0 & -1 \\ -1 & 0 & 0 & 0 \\ 0 & 1 & 0 & 0 \end{bmatrix}^2 = \begin{bmatrix} -1 & 0 & 0 & 0 \\ 0 & -1 & 0 & 0 \\ 0 & 0 & -1 & 0 \\ 0 & 0 & 0 & -1 \end{bmatrix}
$$

(7.9)

Within Clifford algebras, the basis elements always square to either plus unity or minus unity. We see that symmetric and anti-symmetric permutation matrices have the same property. In a manner analogous to the above (7.2) to (7.6), such matrices comprise all the $C_2 \times C_2 \times$... algebras, and so we see that there might be[52] connections between the Clifford algebras and the $C_2 \times C_2 \times$... division algebras.

---

[52] It is not just 'might be'. There are connections. We would not have included this chapter otherwise.

Aside:

Within quantum mechanics, observables are represented as self-adjoint matrices with complex elements (that is Hermitian matrices). Every self-adjoint complex matrix is really a symmetric permutation matrix. To see this, simply use the 2×2 matrix form of the complex numbers in the self-adjoint matrix. Of course, symmetric permutation matrices have real eigenvalues and orthogonal eigenvectors. We can rewrite quantum mechanics using symmetric permutation matrices, and so we have an intimate connection between the $C_2 \times C_2 \times ...$ finite groups and quantum physics.

Chapter 8

Conjugation

Within conventional Clifford algebra, we have three conjugation like operations. In $Cl_{2,0}$, we have[53]:

$$
\text{Grade involution} \quad
\begin{cases}
a + b\vec{e_1} + c\vec{e_2} + d\vec{e_{12}} \rightarrow \\
u = a - b\vec{e_1} - c\vec{e_2} + d\vec{e_{12}}
\end{cases}
$$

$$
\text{Reversion} \quad
\begin{cases}
a + b\vec{e_1} + c\vec{e_2} + d\vec{e_{12}} \rightarrow \\
u = a + b\vec{e_1} + c\vec{e_2} - d\vec{e_{12}}
\end{cases}
\qquad (8.1)
$$

$$
\text{Clifford conjugation} \quad
\begin{cases}
a + b\vec{e_1} + c\vec{e_2} + d\vec{e_{12}} \rightarrow \\
\bar{u} = a - b\vec{e_1} - c\vec{e_2} - d\vec{e_{12}}
\end{cases}
$$

In $Cl_{3,0}$, these conjugations are[54]:

$$
\text{Grade involution} \quad
\begin{cases}
a + b_i\vec{e_i} + c_{jk}\vec{e_{jk}} + d\vec{e_{123}} \rightarrow \\
u = a - b_i\vec{e_i} + c_{jk}\vec{e_{jk}} - d\vec{e_{123}}
\end{cases}
$$

$$
\text{Reversion} \quad
\begin{cases}
a + b_i\vec{e_i} + c_{jk}\vec{e_{jk}} + d\vec{e_{123}} \rightarrow \\
u = a + b_i\vec{e_i} - c_{jk}\vec{e_{jk}} - d\vec{e_{123}}
\end{cases}
\qquad (8.2)
$$

$$
\text{Clifford conjugation} \quad
\begin{cases}
a + b_i\vec{e_i} + c_{jk}\vec{e_{jk}} + d\vec{e_{123}} \rightarrow \\
\bar{u} = a - b_i\vec{e_i} - c_{jk}\vec{e_{jk}} + d\vec{e_{123}}
\end{cases}
$$

In $Cl_{4,0}$, these conjugations are[55]:

---

[53] Lounesto: Page 29
[54] Lounesto: Page 56
[55] Lounesto: Page 86

Grade inversion
$$\begin{cases} a+b_i\overrightarrow{e_i}+c_{jk}\overrightarrow{e_{jk}}+d_{lmn}\overrightarrow{e_{lmn}}+f\overrightarrow{e_{1234}} \rightarrow \\ u=a-b_i\overrightarrow{e_i}+c_{jk}\overrightarrow{e_{jk}}-d_{lmn}\overrightarrow{e_{lmn}}+f\overrightarrow{e_{1234}} \end{cases}$$

Reversion
$$\begin{cases} a+b_i\overrightarrow{e_i}+c_{jk}\overrightarrow{e_{jk}}+d_{lmn}\overrightarrow{e_{lmn}}+f\overrightarrow{e_{1234}} \rightarrow \\ u=a+b_i\overrightarrow{e_i}-c_{jk}\overrightarrow{e_{jk}}-d_{lmn}\overrightarrow{e_{lmn}}+f\overrightarrow{e_{1234}} \end{cases} \quad (8.3)$$

Clifford conjugation
$$\begin{cases} a+b_i\overrightarrow{e_i}+c_{jk}\overrightarrow{e_{jk}}+d_{lmn}\overrightarrow{e_{lmn}}+f\overrightarrow{e_{1234}} \rightarrow \\ \overline{u}=a-b_i\overrightarrow{e_i}-c_{jk}\overrightarrow{e_{jk}}+d_{lmn}\overrightarrow{e_{lmn}}+f\overrightarrow{e_{1234}} \end{cases}$$

Why all these tyes of conjugation? It is because the Clifford algebras have squares of plus unity as well as square roots of minus unity. Within the division algebras in general, we have only one type of conjugation, and so the whole idea of conjugation is simpler within the division algebras than it is in the Clifford algebras.

Conjugation is just reverse rotation:

Conjugation in the division algebras is reverse rotation. Within the complex numbers, $\mathbb{C}$, we can see that the conjugate of a complex number at angle plus θ with the real axis is the complex number at angle minus θ with the real axis. The two rotations 'cancel' each other and put us on to the real axis.

In all division algebras, conjugation is just changing the sign of every imaginary variable in the polar form of the algebra; we simply reverse the rotation. Things are not so simple in the Cartesian form of the algebra.

Being familiar with the conjugate of a complex number, $\mathbb{C}$, and the conjugate of a quaternion, $\mathbb{H}$, the reader might think that division algebra conjugation of the Cartesian form of the algebra can be achieved by changing the sign of every imaginary variable in the Cartesian form of the algebra; it cannot. In general, conjugation of the Cartesian form of a division algebra is much more complicated than just reversing the signs of the Cartesian imaginary variables.

Conjugation is the same as an equal rotation in the opposite direction. Of course, rotation within division algebra spaces is multi-dimensional rotation expressed as a single rotation matrix. Rotation in division algebra spaces is not the familiar set of 2-dimensional rotations we observe in our space-time and express as a set of six 4×4 matrices containing 2-dimensional trigonometric functions and some zeros and ones, except in the 2-dimensional case, of course. An example of division algebra conjugation is the quaternions in which the inverse of a 'clockwise' rotation through the angle, say $\{\theta, \kappa, \xi\}$, is an 'anti-clockwise' rotation through the angle, say $\{-\theta, -\kappa, -\xi\}$.

The conjugate of division algebras in Cartesian form:
The product of a division algebra matrix and its conjugate must be a multiple of the identity (a real number). The conjugate matrix of a division algebra matrix, A, must therefore be the adjoint of that matrix, $adj(A)$. We have the definition of the adjoint of a matrix:

$$\left(\text{Adjoint}(A) \right) * A = I.Det(A) \qquad (8.4)$$

And so, we conjugate Cartesian form of a division algebra matrix, not be reversing the signs of the imaginary variables but by taking the adjoint of that matrix. It is not quite that simple because we then need to divide by some power of the determinant.

In 2-dimensions, we have:

$$Adj\left(\begin{bmatrix} a & b \\ -b & a \end{bmatrix} \right) = \begin{bmatrix} a & -b \\ b & a \end{bmatrix} \quad \& \quad Adj\left(\begin{bmatrix} t & z \\ z & t \end{bmatrix} \right) = \begin{bmatrix} t & -z \\ -z & t \end{bmatrix} \qquad (8.5)$$

These are familiar sign reversal conjugates. In 4-dimensions, we have the quaternions:

$$Adj\left(\begin{bmatrix} a & b & c & d \\ -b & a & -d & c \\ -c & d & a & -b \\ -d & -c & b & a \end{bmatrix}\right) = \left(a^2 + b^2 + c^2 + d^2\right)\begin{bmatrix} a & -b & -c & -d \\ b & a & d & -c \\ c & -d & a & b \\ d & c & -b & a \end{bmatrix}$$

$$det\left(\begin{bmatrix} a & b & c & d \\ -b & a & -d & c \\ -c & d & a & -b \\ -d & -c & b & a \end{bmatrix}\right) = \left(a^2 + b^2 + c^2 + d^2\right)^2$$

(8.6)

We see that we must divide the adjoint quaternion by the square root of the determinant to get the conjugate. Since the rotation matrix is unit length elements of the algebra (determinant is unity), this is no problem to us in the polar form of the algebra. We also have the A_3 algebra:

$$Adj\left(\begin{bmatrix} a & b & c & d \\ b & a & d & c \\ c & -d & a & -b \\ -d & c & -b & a \end{bmatrix}\right) = \left(a^2 - b^2 - c^2 + d^2\right)\begin{bmatrix} a & -b & -c & -d \\ -b & a & -d & -c \\ -c & d & a & b \\ d & -c & b & a \end{bmatrix}$$

$$det\left(\begin{bmatrix} a & b & c & d \\ b & a & d & c \\ c & -d & a & -b \\ -d & c & -b & a \end{bmatrix}\right) = \left(a^2 - b^2 - c^2 + d^2\right)^2$$

(8.7)

Wherein we also divide the adjoint by the square root of the determinant to get the conjugate.

In 8-dimensions, we have the adjoint of the $\left\{1, \sqrt{+1}, 6\sqrt{-1}\right\} = DQ$ algebra, see (13.1), is:

$$adj(DQ)_{[1,1]} = \left[a\left(\begin{array}{c} a^2 + b^2 + c^2 + d^2 \\ -e^2 + f^2 + g^2 + h^2 \end{array} \right) \atop -2e(bf + cg + dh) \right] \sqrt{\det(DQ)}$$

$$adj(DQ)_{[1,2]} = \left[b\left(\begin{array}{c} a^2 + b^2 + c^2 + d^2 \\ +e^2 - f^2 + g^2 + h^2 \end{array} \right) \atop -2f(ae + cg + dh) \right] \sqrt{\det(DQ)}$$

(8.8)

$$adj(DQ)_{[1,3]} = \left[c\left(\begin{array}{c} a^2 + b^2 + c^2 + d^2 \\ +e^2 + f^2 - g^2 + h^2 \end{array} \right) \atop -2g(ae + bf + dh) \right] \sqrt{\det(DQ)}$$

$$adj(DQ)_{[1,4]} = \left[d\left(\begin{array}{c} a^2 + b^2 + c^2 + d^2 \\ +e^2 + f^2 + g^2 - h^2 \end{array} \right) \atop -2h(ae + bf + cg) \right] \sqrt{\det(DQ)}$$

(8.9)

$$adj(DQ)_{[1,5]} = \left[e\left(\begin{array}{c} a^2 - b^2 - c^2 - d^2 \\ -e^2 - f^2 - g^2 - h^2 \end{array} \right) \atop +2a(bf + cg + dh) \right] \sqrt{\det(DQ)}$$

$$adj(DQ)_{[1,6]} = \left[f\left(\begin{array}{c} a^2 - b^2 + c^2 + d^2 \\ +e^2 + f^2 + g^2 + h^2 \end{array} \right) \atop -2b(ae + cg + dh) \right] \sqrt{\det(DQ)}$$

(8.10)

$$adj(DQ)_{[1,7]} = \left[g\left(\begin{array}{c} a^2 + b^2 - c^2 + d^2 \\ +e^2 + f^2 + g^2 + h^2 \end{array} \right) \atop -2c(ae + bf + dh) \right] \sqrt{\det(DQ)}$$

$$adj(DQ)_{[1,8]} = \left[f\left(\begin{array}{c} a^2 + b^2 + c^2 - d^2 \\ +e^2 + f^2 + g^2 + h^2 \end{array} \right) \atop -2d(ae + bf + cg) \right] \sqrt{\det(DQ)}$$

(8.11)

For presentational ease, we have given only the top row of the matrix. Note that the e element corresponds to the commutative 3-vector $\overrightarrow{e_{123}}$ in the 8-dimensional Clifford algebras. The determinant of this 8-dimensional algebra is:

$$\det(DQ) = \left[\left(\begin{array}{c}(a+e)^2 + (b+f)^2 \\ +(c+g)^2 + (d+h)^2\end{array}\right)\left(\begin{array}{c}(a-e)^2 + (b-f)^2 \\ +(c-g)^2 + (d-h)^2\end{array}\right)\right]^2 \quad (8.12)$$

It is intriguing that we have similarities to the 4-dimensional distance functions but with variables doubled together – more upon this later.

Conjugating rotation matrices in division algebras:
The example of conjugation of the 8-dimensional algebra above (8.8) to (8.11) seems very complicated compared to the 2-dimensional or 4-dimensional algebras. The reader whose concentration has lapsed might think this would be indicative of a complicated rotation, but it is not indicative of a complicated rotation. We demonstrate with a 3-dimensional division algebra based on the finite group C_3. We have:

$$\exp\left(\begin{bmatrix} a & b & c \\ c & a & b \\ b & c & a \end{bmatrix}\right) = \begin{bmatrix} e^a & 0 & 0 \\ 0 & e^a & 0 \\ 0 & 0 & e^a \end{bmatrix}\begin{bmatrix} v_A(b,c) & v_B(b,c) & v_C(b,c) \\ v_C(b,c) & v_A(b,c) & v_B(b,c) \\ v_B(b,c) & v_C(b,c) & v_A(b,c) \end{bmatrix}$$

$$(8.13)$$

Where:

$$v_A(b,c) = \frac{1}{3}\left(e^{(b+c)} + 2e^{-\left(\frac{b+c}{2}\right)}\cos\left(\frac{\sqrt{3}}{2}(b-c)\right)\right) \quad (8.14)$$

$$v_B(b,c) = \frac{1}{3}\left(e^{(b+c)} + e^{-\left(\frac{b+c}{2}\right)}\left(\sqrt{3}\sin\left(\frac{\sqrt{3}}{2}(b-c)\right) - \cos\left(\frac{\sqrt{3}}{2}(b-c)\right)\right)\right)$$

$$(8.15)$$

$$v_C(b,c) = \frac{1}{3}\left(e^{(b+c)} - e^{-\left(\frac{b+c}{2}\right)} \left(\sqrt{3}\sin\left(\frac{\sqrt{3}}{2}(b-c) \right) + \cos\left(\frac{\sqrt{3}}{2}(b-c) \right) \right) \right)$$

(8.16)

These,(8.14), (8.15), (8.16), are a set 3-dimensional trigonometric functions. They are each a projection on to an axis of the algebra from the 3-dimensional unit sphere[56] and they appear in a rotation matrix. Since, in general, the determinant of the exponential of a matrix with zero trace is unity, we have:

$$\det\left(\begin{bmatrix} v_A(b,c) & v_B(b,c) & v_C(b,c) \\ v_C(b,c) & v_A(b,c) & v_B(b,c) \\ v_B(b,c) & v_C(b,c) & v_A(b,c) \end{bmatrix} \right) = 1$$

$$v_A^3 + v_B^3 + v_C^3 - 3v_A v_B v_C = 1$$

(8.17)

as direct calculation will verify.

Conjugation within the Cartesian form of this algebra is:

$$adj\left(\begin{bmatrix} a & b & c \\ c & a & b \\ b & c & a \end{bmatrix} \right) = \begin{bmatrix} a^2-bc & c^2-ab & b^2-ac \\ b^2-ac & a^2-bc & c^2-ab \\ c^2-ab & b^2-ac & a^2-bc \end{bmatrix}$$

(8.18)

However, the conjugate in the polar form is obtained by simply reversing the signs of the angle variables:

$$conj\left(\begin{bmatrix} v_A(b,c) & v_B(b,c) & v_C(b,c) \\ v_C(b,c) & v_A(b,c) & v_B(b,c) \\ v_B(b,c) & v_C(b,c) & v_A(b,c) \end{bmatrix} \right)$$

$$= \begin{bmatrix} v_A(-b,-c) & v_B(-b,-c) & v_C(-b,-c) \\ v_C(-b,-c) & v_A(-b,-c) & v_B(-b,-c) \\ v_B(-b,-c) & v_C(-b,-c) & v_A(-b,-c) \end{bmatrix}$$

(8.19)

---

[56] That is sphere in a very general sense that includes non-euclidean spheres.

We can verify this either by directly calculating the adjoint of the rotation matrix in (8.13) or by multiplying together the rotation matrix and its conjugate:

$$PROD \left(\begin{bmatrix} v_A(b,c) & v_B(b,c) & v_C(b,c) \\ v_C(b,c) & v_A(b,c) & v_B(b,c) \\ v_B(b,c) & v_C(b,c) & v_A(b,c) \end{bmatrix} \begin{bmatrix} v_A(-b,-c) & v_B(-b,-c) & v_C(-b,-c) \\ v_C(-b,-c) & v_A(-b,-c) & v_B(-b,-c) \\ v_B(-b,-c) & v_C(-b,-c) & v_A(-b,-c) \end{bmatrix} \right) = \begin{bmatrix} 1 & 0 & 0 \\ 0 & 1 & 0 \\ 0 & 0 & 1 \end{bmatrix} \quad (8.20)$$

We see in this 3-dimensional example that conjugation of the polar form is simply a reverse rotation. The seeming complexity of the Cartesian form of the conjugation is secondary to the fact that a rotation through a given *n*-dimensional angle followed by an equal rotation in the reverse direction gives the identity.

Summary:
Within division algebras, conjugation is no more than reverse rotation. The three types of Clifford algebra conjugation in general are related to such reverse rotation in a very complicated way. We don't need the complications, and so we have no use of the three Clifford conjugations.

Chapter 9

The $C_2 \times C_2$ Division Algebras

There are sixteen[57] 4-dimensional division algebras that derive from the order four finite group $C_2 \times C_2$, however there is much algebraic isomorphism between these sixteen algebras and there are only four algebraically distinct 4-dimensional $C_2 \times C_2$ division algebras. These four algebras are $\{\mathbb{H}, A_1, A_2, A_3\}$. These four distinct algebras have the following ratios of square roots of plus unity and square roots of minus unity (that is symmetric permutation matrices and anti-symmetric permutation matrices or 2-dimensional sub-algebras):

$$
\begin{array}{ccc}
 & \sqrt{+1} & \sqrt{-1} \\
A_1 & 3 & 0 \\
A_2 & 1 & 2 \\
A_3 & 2 & 1 \\
\mathbb{H} & 0 & 3
\end{array}
\qquad (9.1)
$$

Of these four distinct 4-dimensional division algebras, only two $\{\mathbb{H}, A_3\}$ are non-commutative. In more than two dimensions, Clifford algebras are always non-commutative, and so we will ignore the $\{A_1, A_2\}$ commutative division algebras[58] for the present. This leaves only the two non-commutative 4-dimensional division algebras; these

---

[57] Although the derivation of these algebras is systematically dealt with elsewhere (see prerequisites), we can form them by simply combining together 4×4 symmetric and anti-symmetric permutation matrices in such a way to give a multiplicatively closed matrix form and then taking the exponential to give the polar form of the algebra.

[58] In more than two dimensions, it seems that the physics of the observed universe is derived from only non-commutative algebras.

are the $\{A_3, \mathbb{H}\}$ algebras. The numbers of square roots of minus unity and square roots of plus unity (the relative numbers of different 2-dimensional sub-algebras) within the three 4-dimensional Clifford algebras match the two non-commutative 4-dimensional division algebras. We can associate the three 4-dimensional Clifford algebras with them:

$$Cl_{2,0} \sim Cl_{1,1} \sim \left\{ \begin{array}{l} 1, \sqrt{+1}, \\ \sqrt{+1}, \sqrt{-1} \end{array} \right\} \sim A_3 \sim \exp\left(\begin{bmatrix} a & b & c & d \\ b & a & d & c \\ c & -d & a & -b \\ -d & c & -b & a \end{bmatrix} \right) \quad (9.2)$$

$$Cl_{0,2} \sim \left\{ \begin{array}{l} 1, \sqrt{-1}, \\ \sqrt{-1}, \sqrt{-1} \end{array} \right\} \sim \mathbb{H} \sim \exp\left(\begin{bmatrix} a & b & c & d \\ -b & a & -d & c \\ -c & d & a & -b \\ -d & -c & b & a \end{bmatrix} \right) \quad (9.3)$$

Algebraically isomorphic Clifford algebras:
We see that, when we consider only the relative numbers of 2-dimensional sub-algebras, the two Clifford algebras $Cl_{2,0}$ & $Cl_{1,1}$ are the same algebra. Within the $Cl_{2,0}$ algebra, the two basis vectors are associated with square roots of plus unity and the basis bi-vector is associated with the square root of minus unity. Within the $Cl_{1,1}$ algebra, one basis vector and the basis bi-vector are each associated with square roots of plus unity and one basis vector is associated with the square root of minus unity.

$$
\begin{aligned}
Cl_{2,0} &\quad : \quad 1, \ \vec{e_1} = \sqrt{+1}, \ \vec{e_2} = \sqrt{+1}, \ \vec{e_{12}} = \sqrt{-1} \\
Cl_{0,2} &\quad : \quad 1, \ \vec{e_1} = \sqrt{-1}, \ \vec{e_2} = \sqrt{-1}, \ \vec{e_{12}} = \sqrt{-1} \quad (9.4) \\
Cl_{1,1} &\quad : \quad 1, \ \vec{e_1} = \sqrt{+1}, \ \vec{e_2} = \sqrt{-1}, \ \vec{e_{12}} = \sqrt{+1}
\end{aligned}
$$

We see that in $Cl_{2,0}$ & $Cl_{1,1}$ a vector has swapped function with a bi-vector and so the two Clifford algebras $Cl_{2,0}$ & $Cl_{1,1}$ are different only if we distinguish between basis vectors and basis bi-vectors. The only difference between a basis bi-vector and a basis vector is in the conventional interpretation of these objects as representing an oriented area and an oriented length respectively. Mathematically, there is no difference between these two objects. We, as Georgi[59], hold the view that this area/length interpretation is not essential to the Clifford algebras, and so we hold the view that the Clifford algebras $Cl_{2,0}$ & $Cl_{1,1}$ are the same algebra. Technically, they are algebraically isomorphic to each other[60].

Clifford algebras and division algebras in four dimensions:
The four basis elements of the $Cl_{2,0}$ Clifford algebra have the same algebraic properties as the variables (when set to unity) in the A_3 algebras. We take the *SSA* A_3 algebra as an example:

$$SSA = \exp\left(\begin{bmatrix} a & b & c & d \\ b & a & d & c \\ c & -d & a & -b \\ -d & c & -b & a \end{bmatrix}\right) \tag{9.5}$$

The two vectors in $Cl_{2,0}$ are square roots of plus unity which correspond to two symmetric matrix variables, $\{b,c\}$, and the bi-vector in $Cl_{2,0}$ is a square root of minus unity which corresponds to an anti-symmetric matrix variable, $\{d\}$.

---

[59] See references at the start of this book.
[60] Even more technically, algebraic isomorphism is defined between only division algebras, and so Clifford algebras cannot be algebraically isomorphic to each other, but you get the idea.

Aside:

It is technically incorrect to say the $Cl_{2,0}$ algebra is isomorphic to the *SSA* algebra because the polar form of the *SSA* algebra given above is a division algebra whilst a Clifford algebra is not a division algebra.

Separating the algebra into separate variables, we see that the matrix product of the two symmetric variables $\{b,c\}$ is an anti-symmetric variable $\{d\}$ and that the sign of the anti-symmetric variable is changed with the order of multiplication:

$$
\begin{bmatrix} 0 & b & 0 & 0 \\ b & 0 & 0 & 0 \\ 0 & 0 & 0 & -b \\ 0 & 0 & -b & 0 \end{bmatrix}
\begin{bmatrix} 0 & 0 & c & 0 \\ 0 & 0 & 0 & c \\ c & 0 & 0 & 0 \\ 0 & c & 0 & 0 \end{bmatrix}
=
\begin{bmatrix} 0 & 0 & 0 & bc \\ 0 & 0 & bc & 0 \\ 0 & -bc & 0 & 0 \\ -bc & 0 & 0 & 0 \end{bmatrix}
$$

$$
\begin{bmatrix} 0 & 0 & c & 0 \\ 0 & 0 & 0 & c \\ c & 0 & 0 & 0 \\ 0 & c & 0 & 0 \end{bmatrix}
\begin{bmatrix} 0 & b & 0 & 0 \\ b & 0 & 0 & 0 \\ 0 & 0 & 0 & -b \\ 0 & 0 & -b & 0 \end{bmatrix}
=
\begin{bmatrix} 0 & 0 & 0 & -bc \\ 0 & 0 & -bc & 0 \\ 0 & bc & 0 & 0 \\ bc & 0 & 0 & 0 \end{bmatrix}
$$

(9.6)

With these relationships in mind, we have:

$$b \equiv \vec{e_1}, \quad c \equiv \vec{e_2}, \quad d \equiv \vec{e_{12}}$$
$$bc = d, \quad cb = -d$$
$$\vec{e_1}\vec{e_2} = -\vec{e_2}\vec{e_1}$$

(9.7)

In matrix form[61], the Clifford algebra $Cl_{2,0}$ is:

---

[61] It is not conventional to write a 2^n dimensional Clifford algebra as a $2^n \times 2^n$ matrix, and so we are introducing a new aspect of the Clifford algebras by so writing them.

$$Cl_{2,0} = \begin{bmatrix} a & b\vec{e_1} & c\vec{e_2} & d\vec{e_{12}} \\ b\vec{e_1} & a & d\vec{e_{12}} & c\vec{e_2} \\ c\vec{e_2} & -d\vec{e_{12}} & a & -b\vec{e_1} \\ -d\vec{e_{12}} & c\vec{e_2} & -b\vec{e_1} & a \end{bmatrix} \qquad (9.8)$$

To be honest, the above matrix ought to not have the vectors and bi-vector within it. The algebraic properties of these objects is encoded within the matrix form, and we really ought to write:

$$Cl_{2,0} = \begin{bmatrix} a & b & c & d \\ b & a & d & c \\ c & -d & a & -b \\ -d & c & -b & a \end{bmatrix} \qquad (9.9)$$

The lack of a multiplicative inverse within the Clifford algebra is now expressed as the possibility of this matrix, (9.9), being singular. The determinant of (9.9) is of the form:

$$\det\left(Cl_{2,0}\right) = \left(a^2 - b^2 - c^2 + d^2\right)^2 \qquad (9.10)$$

We can rid ourselves of these singular matrices by taking the exponential of the matrix thereby forming a A_3 division algebra.

The preservation of the norms $x^2 + y^2$ and $x^2 - y^2$ which led Clifford to these algebras is now expressed as the algebraic fact that:

$$\left(a^2 - b^2 - c^2 + d^2\right)\left(s^2 - t^2 - u^2 + v^2\right) = W^2 - X^2 - Y^2 + Z^2 \qquad (9.11)$$

The form of the norm is maintained under multiplication because the form of the matrix is maintained under multiplication and so the form of the determinant of the matrix is maintained under multiplication. (The lower dimensional cases of the $C_2 \times C_2 \times ...$ algebras are exceptional, and the norm of the higher dimensional $C_2 \times C_2 \times ...$ algebras is not a simple quadratic form.)

Aside:

By definition, quantum gravity will contain commutation relations. Since multiplication properly exists in only division algebras, commutation relations exist in only division algebras. The distance function $t^2 - x^2 - y^2 - z^2$ is not maintained under multiplication, and so there is no division algebra with this distance function and so there is no quantum gravity with this distance function[62].

The Clifford product:

The $Cl_{2,0}$ Clifford product of two 2-dimensional vectors is:

$$\overline{\overline{ab}} = a_1 b_1 + a_2 b_2 + \left(a_1 b_2 - a_2 b_1 \right) \overrightarrow{e_{12}} \tag{9.12}$$

This is the sum of a scalar and a bi-vector. Within the *SSA* algebra, we have:

$$
\begin{bmatrix}
0 & a_1 & a_2 & 0 \\
a_1 & 0 & 0 & a_2 \\
a_2 & 0 & 0 & -a_1 \\
0 & a_2 & -a_1 & 0
\end{bmatrix}
\begin{bmatrix}
0 & b_1 & b_2 & 0 \\
b_1 & 0 & 0 & b_2 \\
b_2 & 0 & 0 & -b_1 \\
0 & b_2 & -b_1 & 0
\end{bmatrix}
$$

$$
=
\begin{bmatrix}
a_1 b_1 + a_2 b_2 & 0 & 0 & a_1 b_2 - a_2 b_1 \\
0 & a_1 b_1 + a_2 b_2 & a_1 b_2 - a_2 b_1 & 0 \\
0 & -\left(a_1 b_2 - a_2 b_1 \right) & a_1 b_1 + a_2 b_2 & 0 \\
-\left(a_1 b_2 - a_2 b_1 \right) & 0 & 0 & a_1 b_1 + a_2 b_2
\end{bmatrix}
\tag{9.13}
$$

The inner product of the two vectors is the real variable, and the wedge product of the two vectors is the anti-symmetric imaginary variable. We also have (in reverse order):

---

[62] This is far from being generally accepted.

$$
\begin{bmatrix}
0 & b_1 & b_2 & 0 \\
b_1 & 0 & 0 & b_2 \\
b_2 & 0 & 0 & -b_1 \\
0 & b_2 & -b_1 & 0
\end{bmatrix}
\begin{bmatrix}
0 & a_1 & a_2 & 0 \\
a_1 & 0 & 0 & a_2 \\
a_2 & 0 & 0 & -a_1 \\
0 & a_2 & -a_1 & 0
\end{bmatrix}
$$

$$
=
\begin{bmatrix}
a_1 b_1 + a_2 b_2 & 0 & 0 & -\left(a_1 b_2 - a_2 b_1\right) \\
0 & a_1 b_1 + a_2 b_2 & -\left(a_1 b_2 - a_2 b_1\right) & 0 \\
0 & a_1 b_2 - a_2 b_1 & a_1 b_1 + a_2 b_2 & 0 \\
a_1 b_2 - a_2 b_1 & 0 & 0 & a_1 b_1 + a_2 b_2
\end{bmatrix}
$$

$$(9.14)$$

The Clifford product of two vectors is then traditionally written as:

$$
\overline{\overline{ab}} = \vec{a} \bullet \vec{b} + \vec{a} \wedge \vec{b}
$$
$$
\overline{\overline{ba}} = \vec{a} \bullet \vec{b} - \vec{a} \wedge \vec{b}
$$

$$(9.15)$$

We see above that the nature of the two matrices (carefully chosen):

$$
\vec{e_1} \equiv
\begin{bmatrix}
0 & b_1 & 0 & 0 \\
b_1 & 0 & 0 & 0 \\
0 & 0 & 0 & -b_1 \\
0 & 0 & -b_1 & 0
\end{bmatrix}
\quad \& \quad
\vec{e_2} \equiv
\begin{bmatrix}
0 & 0 & a_2 & 0 \\
0 & 0 & 0 & a_2 \\
a_2 & 0 & 0 & 0 \\
0 & a_2 & 0 & 0
\end{bmatrix}
$$

$$(9.16)$$

determines the rest of the algebra. Because only two elements of the matrix are needed to determine the whole matrix, the algebra can be generated with just two elements; this is how a 4-dimensional Clifford algebra is generated by only two basis vectors.

We see that the dot product and the wedge product of the two vectors is given in terms of the Clifford product as:

$$
\vec{a} \bullet \vec{b} = \frac{1}{2}\left(\overline{\overline{ab}} + \overline{\overline{ba}}\right)
$$
$$
\vec{a} \wedge \vec{b} = \frac{1}{2}\left(\overline{\overline{ab}} - \overline{\overline{ba}}\right)
$$

$$(9.17)$$

Looking at (9.13) and (9.14) above, we see:

$$\frac{1}{2}\left(\begin{bmatrix} 0 & a_1 & a_2 & 0 \\ a_1 & 0 & 0 & a_2 \\ a_2 & 0 & 0 & -a_1 \\ 0 & a_2 & -a_1 & 0 \end{bmatrix}\begin{bmatrix} 0 & b_1 & b_2 & 0 \\ b_1 & 0 & 0 & b_2 \\ b_2 & 0 & 0 & -b_1 \\ 0 & b_2 & -b_1 & 0 \end{bmatrix}\right.$$
$$\left.+\begin{bmatrix} 0 & b_1 & b_2 & 0 \\ b_1 & 0 & 0 & b_2 \\ b_2 & 0 & 0 & -b_1 \\ 0 & b_2 & -b_1 & 0 \end{bmatrix}\begin{bmatrix} 0 & a_1 & a_2 & 0 \\ a_1 & 0 & 0 & a_2 \\ a_2 & 0 & 0 & -a_1 \\ 0 & a_2 & -a_1 & 0 \end{bmatrix}\right) \qquad (9.18)$$

$$=\begin{bmatrix} a_1b_1+a_2b_2 & 0 & 0 & 0 \\ 0 & a_1b_1+a_2b_2 & 0 & 0 \\ 0 & 0 & a_1b_1+a_2b_2 & 0 \\ 0 & 0 & 0 & a_1b_1+a_2b_2 \end{bmatrix}$$

$$\frac{1}{2}\left(\begin{bmatrix} 0 & a_1 & a_2 & 0 \\ a_1 & 0 & 0 & a_2 \\ a_2 & 0 & 0 & -a_1 \\ 0 & a_2 & -a_1 & 0 \end{bmatrix}\begin{bmatrix} 0 & b_1 & b_2 & 0 \\ b_1 & 0 & 0 & b_2 \\ b_2 & 0 & 0 & -b_1 \\ 0 & b_2 & -b_1 & 0 \end{bmatrix}\right.$$
$$\left.-\begin{bmatrix} 0 & b_1 & b_2 & 0 \\ b_1 & 0 & 0 & b_2 \\ b_2 & 0 & 0 & -b_1 \\ 0 & b_2 & -b_1 & 0 \end{bmatrix}\begin{bmatrix} 0 & a_1 & a_2 & 0 \\ a_1 & 0 & 0 & a_2 \\ a_2 & 0 & 0 & -a_1 \\ 0 & a_2 & -a_1 & 0 \end{bmatrix}\right) \qquad (9.19)$$

$$=\begin{bmatrix} 0 & 0 & 0 & a_1b_2-a_2b_1 \\ 0 & 0 & a_1b_2-a_2b_1 & 0 \\ 0 & -(a_1b_2-a_2b)_1 & 0 & 0 \\ -(a_1b_2-a_2b_1) & 0 & 0 & 0 \end{bmatrix}$$

We see that we have algebraic isomorphism between the Clifford algebra, $Cl_{2,0}$, and the Cartesian form of the *SSA* A_3 algebra. Of course, only the polar form of the *SSA* A_3 algebra is a full-blown division algebra.

In the case of the $Cl_{1,1}$ algebra, we identify:

$$b \equiv \vec{e_1}, \quad c \equiv \vec{e_{12}}, \quad d \equiv \vec{e_2},$$
$$bd = c, \quad db = -c$$

(9.20)

We can therefore write the Clifford algebra $Cl_{1,1}$ as:

$$\begin{bmatrix} a & b\vec{e_1} & c\vec{e_{12}} & d\vec{e_2} \\ b\vec{e_1} & a & d\vec{e_2} & c\vec{e_{12}} \\ c\vec{e_{12}} & -d\vec{e_2} & a & -b\vec{e_1} \\ -d\vec{e_2} & c\vec{e_{12}} & -b\vec{e_1} & a \end{bmatrix}$$

(9.21)

Again, we've been naughty and we've put the basis vectors and basis bi-vector into the matrix where they ought not to be, but we opine that it presents the material more clearly. Taking the exponential of this will rid us of the singular matrices and thereby give a A_3 division algebra. We will then have turned a Clifford algebra into a division algebra.

How to turn a Clifford algebra into a division algebra:
1) Write the 2^n dimensional Clifford algebra as a $2^n \times 2^n$ matrix of symmetric and anti-symmetric variables which is of multiplicatively closed form.
2) Take the exponential of that matrix.

The general element:
Conventionally, the general element of the $Cl_{2,0}$ algebra is written as:

$$u = u_0(scalar) + u_1(vector) + u_2(bi-vector)$$

(9.22)

Because both $\{\vec{e_1}, \vec{e_2}\}$ are interpreted to be basis vectors in $\mathbb{R}^2$, they are added together to form a single vector. The matrix presentation does not really allow this, and so we take the view that this adding together of basis vectors is interpretive only[63].

Sub-algebras:

When written in matrix form, the division algebras reveal a subtle difference of understanding regarding the nature of sub-algebras. The quaternions are a good example. In the non-matrix notation:

$$Q = a + \hat{i}b + jc + kd \qquad (9.23)$$

Setting $c = d = 0$ gives:

$$a + \hat{i}b \qquad (9.24)$$

Which is 'obviously' a complex number, $\in \mathbb{C}$, and so we say that the complex numbers are a 2-dimensional sub-algebra of the quaternions. This is not quite true, and the notation is deceiving.

In matrix notation:

$$\begin{bmatrix} a & b & 0 & 0 \\ -b & a & 0 & 0 \\ 0 & 0 & a & -b \\ 0 & 0 & b & a \end{bmatrix} \neq \begin{bmatrix} a & b \\ -b & a \end{bmatrix} \qquad (9.25)$$

The complex numbers are a 2-dimensional algebra. The quaternions with two zero imaginary elements are still a 4-dimensional algebra. The 2×2 and the 4×4 forms above, (9.25), are algebraically isomorphic, but they are not identical. This subtle difference is very important in physics. It is the difference between a double covering spinor space and a 'normal' space. Further, Lorentz transformations of electromagnetic fields are done with 2×2 matrices. If we attempt to Lorentz transform with 4×4 matrices, we get the wrong answers. Also, in our 4-dimensional space-time, I can rotate in a 2-dimensional

---

[63] Actually, we think it is wrong, but politeness precludes our direct expression of this opinion. Tactfully, we will not tell anyone.

plane without affecting anything perpendicular to that plane because my rotation is described by a 2×2 rotation matrix; if my rotation was described by a 4×4 rotation matrix, things perpendicular to the plane of rotation would be affected by my rotation. A great deal of physics would be changed if the complex numbers, $\mathbb{C}$, were a 'honest' sub-algebra of the quaternions, $\mathbb{H}$ in the way they are usually presumed to be.

Within $Cl_{2,0}$, there is a 2-dimensional sub-algebra that is isomorphic to the complex numbers, $\mathbb{C}$. This is not the only 2-dimensional sub-algebra within $Cl_{2,0}$, but the other sub-algebras are isomorphic to the hyperbolic complex numbers of which many mathematicians are unaware. So it is that these other sub-algebras are often missed in conventional treatments of this area of mathematics; we will include them. We list the 2-dimensional sub-algebras of the 4-dimensional Clifford algebras. There is one 2-dimensional sub-algebra for each basis element of the Clifford algebra except the scalar. Simply pairing any basis element with the scalar produces the 2-dimensional sub-algebra.

4-dim Clifford algebras	$1, \sqrt{-1}$	$1, \sqrt{+1}$
$Cl_{2,0}$	$\{1, \overrightarrow{e_{12}}\}$	$\{1, \overrightarrow{e_1}\}, \{1, \overrightarrow{e_2}\}$
$Cl_{1,1}$	$\{1, \overrightarrow{e_2}\}$	$\{1, \overrightarrow{e_1}\}, \{1, \overrightarrow{e_{12}}\}$
$Cl_{0,2}$	$\{1, \overrightarrow{e_1}\}, \{1, \overrightarrow{e_2}\}, \{1, \overrightarrow{e_{12}}\}$	0

Even sub-algebras:

We see that the even sub-algebra, $\{1, \overrightarrow{e_{12}}\}$, of both $Cl_{2,0}$ and $Cl_{0,2}$ is the complex numbers, $\mathbb{C}$. The unit length elements of these are the group $spin(2) \cong U(1)$ which is conventionally sometimes[64] seen as a

---

[64] We have written of this 'sometimes' ambiguity above.

set of normalised spinors. We see that the even sub-algebra of $Cl_{1,1}$ is the hyperbolic complex numbers, $\mathbb{S}$. The unit length elements of these are a group which is conventionally sometimes seen as a set of normalised spinors in 2-dimensional space-time, $spin(1,1)$.

The centre of the algebra:

The centre of the *SSA* algebra is just the real scalar. The same is true of all the 4-dimensional Clifford algebras.

Quaternions and $Cl_{0,2}$:

What we have done above with the $Cl_{2,0}$ algebra and the *SSA* A_3 algebra, we can do with the $Cl_{0,2}$ algebra and the quaternions. It is a standard result that the Clifford algebra $Cl_{0,2}$ is isomorphic to the quaternions[65]. We have the Clifford product:

$$\overline{\overline{ab}} = -\left(a_1 b_1 + a_2 b_2\right) + \left(a_1 b_2 - a_2 b_1\right)\overrightarrow{e_{12}} \qquad (9.26)$$

This is the sum of a scalar and a bi-vector. Within the quaternion algebra, we have:

$$
\begin{bmatrix}
0 & a_1 & a_2 & 0 \\
-a_1 & 0 & 0 & a_2 \\
-a_2 & 0 & 0 & -a_1 \\
0 & -a_2 & a_1 & 0
\end{bmatrix}
\begin{bmatrix}
0 & b_1 & b_2 & 0 \\
-b_1 & 0 & 0 & b_2 \\
-b_2 & 0 & 0 & -b_1 \\
0 & -b_2 & b_1 & 0
\end{bmatrix}
$$

$$
=
\begin{bmatrix}
-a_1 b_1 - a_2 b_2 & 0 & 0 & a_1 b_2 - a_2 b_1 \\
0 & -a_1 b_1 - a_2 b_2 & -\left(a_1 b_2 - a_2 b_1\right) & 0 \\
0 & a_1 b_2 - a_2 b_1 & -a_1 b_1 - a_2 b_2 & 0 \\
-\left(a_1 b_2 - a_2 b_1\right) & 0 & 0 & -a_1 b_1 - a_2 b_2
\end{bmatrix}
$$

$$(9.27)$$

---

[65] Lounesto: page 205.

The quaternions are themselves an even sub-algebra of $Cl_{3,0}$, and so the unit length quaternions are conventionally seen as normalised spinors and form a spin group, $spin(3)$.

Aside:
We reiterate: The conventional confusion between whether a spin group is within a whole Clifford algebra or in only an even sub-algebra of a Clifford algebra stems from the failure to realise that the every Clifford algebra is isomorphic to the non-polar form of a division algebra. Convention knows of only the complex numbers and the quaternions and so finds spin groups in only these types of algebras or sub-algebras.

Right-chiral quaternions:
The $C_2 \times C_2$ group contains two quaternion algebras, the left-chiral quaternions and the right-chiral quaternions. Only the left-chiral quaternions emerge from the appropriate Clifford algebras. The right-chiral quaternions are a reversed form of the quaternions, and they would emerge from the Clifford algebras instead of the quaternions if we defined $\overrightarrow{e_1 e_2} = -\overrightarrow{e_{12}}$ rather than the accepted definition[66]. The same is true of the A_3 right-chiral algebras. The $C_2 \times C_2$ group approach does the job completely whereas the conventional Clifford algebra approach misses the right-chiral algebras.

Summary:
We have found a match between the 4-dimensional Clifford algebras and the two non-commutative 4-dimensional division algebras the quaternions and the A_3 algebras. We have taken $Cl_{2,0}$ to be the same

---

[66] $\overrightarrow{e_{12}}$ is an object in its own right. It is not the same object as $\overrightarrow{e_1 e_2}$ except by definition.

as $Cl_{1,1}$ because they have the same number of square roots of minus unity and the same number of square roots of plus unity[67]. If we reject the interpretational difference between bi-vectors and vectors, this isomorphism of the two algebras is inevitable. Interpretation is not mathematics. We see that, in four dimensions, there are no unmatched Clifford algebras, and there are no unmatched division algebras (except the commutative algebras and the right-chiral algebras mentioned above).

The traditional Clifford product of vectors is just matrix multiplication with zero scalar and one other zero element, but we can simply expand it to be the product of all four elements if we wish. At a 4-dimensional level, within the division algebras, we have everything that we have in the corresponding Clifford algebras except the length/area interpretation, but, because we are working in division algebras, we have more; for example, taking the exponential of the 4×4 *SSA* A_3 matrix form will give a 4×4 rotation matrix containing 4-dimensional trigonometric functions. We do not have to view bi-vectors as rotations.

People who are wise in the lore of the $C_2 \times C_2$ algebras know that the electromagnetic tensor and the Maxwell equations, the Lorentz Lie group $SO(3,1)$, the absence of anti-matter in our classical universe, the distance function of space-time $d^2 = t^2 - x^2 - y^2 - z^2$ and with that the metric tensor of space-time leading to the field equations of general relativity are within these algebras, and so we have a connection between the Clifford algebras and physics.

A note on spin groups:
In conventional Clifford algebra, a spin group is seen as the 'rotation matrix' of the even sub-algebra of the Clifford algebra. In this book, we see a spin group as the rotation matrix of the whole algebra. Since

---

[67] Lounesto page 212. "The Clifford algebras $Cl_{2,0}$ and $Cl_{1,1}$ are isomorphic as associative algebras but not isomorphic as quadratic algebras."

sub-algebras of both Clifford algebras and division algebras are also Clifford algebras and division algebras respectively, we get the same set of spin groups. It is just that we associate each spin group with a Clifford algebra other than the Clifford algebra with which it is conventionally associated. We are tidier.

It's not really Clifford algebra, but...:
If we take the relations:

$$\vec{e_1} = \sqrt{+1}, \quad \vec{e_2} = \sqrt{+1}, \quad \vec{e_1}\vec{e_2} = \vec{e_2}\vec{e_1} \qquad (9.28)$$

in which the vectors are commutative, we get the commutative A_1 division algebra. If we take the relations:

$$\vec{e_1} = \sqrt{-1}, \quad \vec{e_2} = \sqrt{-1}, \quad \vec{e_1}\vec{e_2} = \vec{e_2}\vec{e_1} \qquad (9.29)$$

we get the commutative A_2 division algebra.

We see that we do not need non-commutative vectors to generate a Clifford like algebra. The other 4-dimensional division algebras, the algebras that derive from the C_4 finite group, are based upon 4th roots of plus unity or minus unity. These are part of the complete picture, but they are not entities in conventional Clifford algebra. Not only are we tidier, we are also comprehensive.

Chapter 10

Our Space-time from Division Algebras

Hermann Gunther Grassmann (1809-1877) formulated vectors as part of an attempt to describe the geometrical nature of space. Clifford took Grasmann's vectors and formulated Clifford algebra as the algebra of the geometry of space. We are rewriting Clifford algebra. Does our rewrite solve the problem first approached by Grassmann? Yes it does, but by a route completely different from what we might have expected.

From division algebras to our space-time:

There are six A_3 algebras within the finite group $C_2 \times C_2$. They are algebraically isomorphic to each other, but they are written in different bases. Within mathematics, we cannot add elements of an algebra written in one basis to elements of that same algebra written in a different basis. However, we can think of these six division algebra spaces as fields and we can simply superimpose these fields on top of each other; the result is simply adding the six algebras together.

When we do this, because we cannot add elements of an algebra that are written in different bases, we break the algebras, and so we break the rotation matrices and other aspects of those algebras. If we have broken the algebra, we have broken the multiplication operation of that algebra.

Proper multiplication exists in only division algebras. Our 4-dimensional space-time is not a division algebra space, and so we cannot multiply within our 4-dimensional space-time. Without multiplication, we have no commutation relations and no imaginary square roots of plus unity or minus unity in our 4-dimensional space-time, but we are left with ordered sets of four numbers specifying each position in our 4-dimensional space-time.

Our first step is to recognise that four real numbers alone are no more than a continuous manifold. Each point has four numbers associated with it, and so we have the concept of distinct points continuously connected together.

Within a manifold, we have no sense of distance between two points in the manifold. We calculate a distance between two points by feeding the two sets of four numbers specifying the two points into a distance function, but there is an infinitude of four variable possible distance functions. Unless we know which four variable function to choose as our distance function, the concept of distance between two points is meaningless. How are we to choose the distance function of a manifold?

The distance function of the space-time in which we sit:
We sum of the six individual distances through each A_3 space to form an expectation distance function:

$$SUM \begin{cases} dist^2 = t^2 - x^2 - y^2 + z^2 \\ dist^2 = t^2 - x^2 - y^2 + z^2 \\ dist^2 = t^2 - x^2 + y^2 - z^2 \\ dist^2 = t^2 - x^2 + y^2 - z^2 \\ dist^2 = t^2 + x^2 - y^2 - z^2 \\ dist^2 = t^2 + x^2 - y^2 - z^2 \end{cases} = 2\left(3t^2 - x^2 - y^2 - z^2\right) \quad (10.1)$$

The 2 is just a doubling up because we have counted both the distance through the left-chiral A_3 spaces and through the right-chiral A_3 spaces, or it is just a factor which we can forget. The 3 gets absorbed in the different units which we use to measure space and time. We have the distance function of our 4-dimensional space-time.

Wow! That is an unexpected development of Clifford algebra.

The affine connection:

We have no concept of parallel transport (affine connection) of a vector through the manifold we have formed by adding the A_3 spaces.

Now that we have a distance function, we know the distance between two points; but by which route through the manifold do we travel that distance. Distance alone does not define a straight line (geodesic).

We allow the direction of a A_3 vector (we might call this the A_3 phase) to vary from point to point over the manifold. This is the standard QFT way of producing the potentials of the electro-weak force, the photon field, and the strong force. But wait, since there is no affine connection in the manifold, that is no sense of parallel directions in the manifold, it is meaningless to say that we allow the direction of the vector to vary over the manifold. There is nothing to which we can compare the direction of the vectors. What are we doing?

In the absence of an affine connection in the underlying manifold, we declare that the A_3 vectors are all parallel; this is the affine connection. The locally varying phase of the A_3 vector induces an affine connection, a sense of direction, into the manifold. The locally varying phase defines parallel transport within the manifold.

Taking all the A_3 vectors to be parallel means that force associated with the locally varying phase of the A_3 algebra will be manifest as the curvature of the manifold – sound familiar; this is general relativity?

Wow! Clifford did not forsee that.

The angles:

We need to add the concept of angle, and with it rotation, to our manifold to form a geometric space.

Superimposition of each of the two 2-dimensional algebras leaves each of them unchanged because there is only one copy of each of them derived from the C_2 group.

The 2-dimensional algebras have distance functions which can be accommodated within the A_3 expectation distance function which is the distance function of our 4-dimensional space-time, and so we have 2-dimensional rotations in our 4-dimensional space-time. We could not accommodate a 3-dimensional rotation in our 4-dimensional space-time because the distance function that a 3-dimensional rotation holds invariant just does not fit into the distance function of our 4-dimensional space-time.

The metric tensor:

By insisting that the A_3 vector maintains the same length at all points in the manifold, we induce the metric tensor on to the manifold. Taking the six mixed double differentials of the metric tensor gives the Riemann curvature tensor; contracting the Riemann curvature tensor and combining it with the metric tensor to form a divergence free tensor produces the Einstein tensor.

The mass energy tensor:

Superimposing the symmetric parts of the $E \& B$ fields, see: (18.11), of the six A_3 algebras produces the mass-energy tensor. Putting this equal to the Einstein tensor gives the field equations of general relativity.

Our 4-dimensional space-time:

There we have the space-time in which we sit – not quite so simple, but there it is. This is a Riemannian space that has been 'derived' by the superimposition of six A_3 division algebras derived from the finite group $C_2 \times C_2$. In modern parlance, we might say that space-time has emerged as an expectation space by breaking the A_3 division algebras.

We see that our 4-dimensional space-time has emerged from a set of Clifford algebras.

The Lorentz group $SO(3,1)$:

Just to be tidy, we look at the Lorentz group. Each of the A_3 algebras has commutation relations within it. Taking the individual variables of all six A_3 algebras will give the Lorentz group, $SO(3,1)$. We tend to think of the Lorentz group as a single Lie algebra, but we now see it as an aggregate of six separate division algebras.

The mass-energy tensor:

When we differentiate an A_3 algebra potential non-commutatively, we get a pair of gravito-electromagnetic fields. Superimposing these fields from all six A_3 algebras leads to a 2nd rank symmetric tensor. We take the symmetric part of this to be the mass-energy tensor.

Summary so far:

Well! We seem to have our 4-dimensional space-time together with the curvature of Einstein's general relativity. We found this space by a route utterly unexpected and different from any understanding of empty space postulated by Grassmann or Clifford. None-the-less, we found this space through the Clifford algebras. What a remarkable twist of fate.

Other spaces:

What can be done with the A_3 algebras can surely be done with all the other types of division algebras. Let us consider a few other types of division algebras.

The $C_2 \times C_2$ group contains sixteen algebras. We have dealt with the eight non-commutative algebras above. The other eight algebras, 6 *of* A_2 & 2 *of* A_1, are commutative. Superimposing the distance functions of these algebras gives the distance functions:

$$A_1 : \quad d^4 = 2 \begin{pmatrix} a^4 + b^4 + c^4 + d^4 \\ -2\left(a^2 b^2 + a^2 c^2 + a^2 d^2 + b^2 c^2 + b^2 d^2 + c^2 d^2\right) \end{pmatrix}$$

$$A_2 : \quad d^4 = 2 \begin{pmatrix} 3\left(a^4 + b^4 + c^4 + d^4\right) \\ +2\left(a^2 b^2 + a^2 c^2 + a^2 d^2 + b^2 c^2 + b^2 d^2 + c^2 d^2\right) \end{pmatrix}$$

$$(10.2)$$

There is only one other order four group, C_4. This group contains eight division algebras of two distinct types. Superimposing the distance functions of the two sets of four algebras gives the superimposed distance functions:

$$d^4{}_{H-type} = 4\left(\left(a^2 - c^2\right)^2 - \left(b^2 + d^2\right)^2\right)$$

$$d^4{}_{E-type} = 4\left(\left(a^2 + c^2\right)^2 + \left(b^2 + d^2\right)^2\right)$$

$$(10.3)$$

The above, (10.1), (10.2), (10.3), are the only 4-dimensional $\mathbb{R}^n$ type of spaces that can be derived from the division algebras by superimposition. Notice that none of the superimposed distance functions can accommodate rotation. Setting two variables to zero does not always produce a distance function held invariant by a rotation.

The reader should note the pairing of variables in (10.3) and that the distance functions have the general form of the distance functions of the two 2-dimensional algebras. The C_4 algebras have a single C_2 sub-algebra[68]. They come in three forms. When two of the variables are zero, the 4-dimensional distance function must become one of the C_2 2-dimensional distance functions. It is the need of the 4-dimensional algebras to accommodate the 2-dimensional sub-algebras that leads to the pairing of variables we see in (10.3).

---

[68] The finite group C_4 has a single C_2 sub-group.

What about 3-dimensional spaces?:

There are four 3-dimensional algebras that derive from the C_3 group. These four algebras are all commutative. Of these, only two are algebraically isomorphic. This leads to three distance functions:

$$d^3 = a^3 + b^3 + c^3 - 3abc$$

$$d^3 = a^3 - b^3 - c^3 - 3abc \tag{10.4}$$

$$d^3 = SUM \begin{Bmatrix} a^3 - b^3 + c^3 + 3abc \\ a^3 + b^3 - c^3 + 3abc \end{Bmatrix} = 2a^3 + 6abc$$

We have now calculated all possible types of $\mathbb{R}^n$ space in two, three, and four dimensions. The only non-commutative types of $\mathbb{R}^n$ space are those derived by superimposition of the quaternion algebras which give electromagnetism and our space-time which we think gives gravity. We have a good match with observed reality.

Stop Press:

When this book was first published, we did not know that our 4-dimensional space-time was the only Riemannian space which could support a geometrical structure. It has now been established that our 4-dimensional space-time is the only super-position space which has a geometrical structure[69]. Our 4-dimensional space-time is unique.

---

[69] See : Dennis Morris & Sophie Lacson : The Uniqueness of our Space-time

Chapter 11

A Few Bits

Determinism and probabilistic quantum theory:
There is no proper algebraic connection between the division algebra spaces (which we think are quantum spaces) and the classical space of 4-dimensional space-time. Hence, there is no deterministic relationship between these two parts of the universe. The mathematics of a division algebra is algebraic, and so within itself, the physics in a division algebra space is deterministic. Similarly, the tensor type of mathematics within 4-dimensional space-time holds together, and so within itself, the classical physics of 4-dimensional space-time is deterministic. But, between the two types of space, there is no deterministic relationship; there is just smashed algebras. Hence, an observer in 4-dimensional space-time looking at the division algebra spaces would see them to be non-deterministic because she views them through the 'mathematical nonsense' that is the algebras that have been smashed by superimposition. The observer could, at best, produce only a probabilistic theory of the physics in the division algebra spaces. We have a basic tenet of quantum mechanics.

In particular, an observer in our space-time could not see a 4-dimensional quaternion rotation because only 2-dimensional rotations exist in our space-time. Perhaps, instead, the observer would see two 2-dimensional quaternion rotations (double cover spinor rotations). This does seem to be what physicists call a spinor.

Where does time exist?:
We have seen space-time emerge from the superimposition of the A_3 division algebras. Does time exist within a division algebra? We do not know, but it seems that it must be different from time in the space-time in which we sit. Without time, is determinism meaningful? Well

yes, but measured against a different variable. Perhaps the nature of time in a division algebra space is so different from the nature of time in our space that it effectively does not exist to us. Thus we might expect an instantaneous collapse of a spatially extensive wave-function or some other manifestation of non-locality. The weirdness of quantum physics seems to be evaporating before our eyes.

Conclusion:
We get the various types of division algebra spaces directly from the finite groups. We think these division algebra spaces are quantum spaces (spinor spaces) and that they hold quantum physics (more later). We see that there are only a limited number of such quantum spaces in any dimension. We get the $\mathbb{R}^n$ types of space by superimposition of algebraically isomorphic division algebra spaces. We think the $\mathbb{R}^n$ types of space hold classical physics (more later). We see that there are only a limited number of such spaces in any dimension. This is quite a revolution in how we think of empty space.

A different view:
If we examine a lot of electrons to determine their individual spin, we will get a lot of plus ones and a lot of minus ones. The average electron spin, the expectation value, will be zero. If we examine a lot of A_3 algebras to determine their distance functions, we will get lots of all three distance functions as in (10.1). If we take the average of these distance functions, the expectation distance function, we get:

$$\frac{2}{3}\left(3t^2 - x^2 - y^2 - z^2\right) \tag{11.1}$$

Perhaps the process of superimposition is no more than an averaging process.

Chapter 12

The $C_2 \times C_2 \times C_2$ Division Algebras

We notice that we can effectively separate the four different 4-dimensional division algebras by counting the number of symmetric and anti-symmetric variables in each algebra. We have:

$$
\begin{array}{ll}
2 \text{ off} & A_1 \sim 3\sqrt{+1} \\
6 \text{ off} & A_2 \sim 1\sqrt{+1} : 2\sqrt{-1} \\
6 \text{ off} & A_3 \sim 2\sqrt{+1} : 1\sqrt{-1} \\
2 \text{ off} & \mathbb{H} \sim 3\sqrt{-1}
\end{array}
\tag{12.1}
$$

When we count the number of symmetric and anti-symmetric variables, we are effectively counting the relative number of the two 2-dimensional algebras, $\{\mathbb{C}, \mathbb{S}\}$.

The 8-dimensional algebraic matrix form:
The basic matrix form of the 8-dimensional $C_2 \times C_2 \times C_2$ division algebras is:

$$
\exp\left(
\begin{bmatrix}
a & b & c & d & e & f & g & h \\
b & a & d & c & f & e & h & g \\
c & d & a & b & g & h & e & f \\
d & c & b & a & h & g & f & e \\
e & f & g & h & a & b & c & d \\
f & e & h & g & b & a & d & c \\
g & h & e & f & c & d & a & b \\
h & g & f & e & d & c & b & a
\end{bmatrix}
\right)
\tag{12.2}
$$

The above matrix is a division algebra in which there are seven symmetric imaginary variables (square roots of plus unity). It is commutative. Inserting scaling parameters will scatter some minus signs about this matrix to form some anti-symmetric variables (square roots of minus unity) and lead to other algebras. Some of these other algebras are non-commutative. However, if we do not consider commutativity, we can write down a set of relationships between the imaginary variables in the matrix. We have:

$$b^2 = a, \quad c^2 = a, \quad d^2 = a, \quad e^2 = a$$
$$f^2 = a, \quad g^2 = a, \quad h^2 = a$$

(12.3)

Other that sign, these match the seven non-identity basis elements of the 8-dimensional Clifford algebras. We also have relations like:

$$bc = d, \quad bd = c, \quad be = f, \quad bf = e, \quad bg = h, \quad bh = g \quad (12.4)$$

The game is to match these relations between imaginary variables to the relations between the basis elements of a Clifford algebra and thereby identify each of these seven imaginary division algebra variables with a basis element of the appropriate Clifford algebra. It is done by informed guesswork[70], and we will do it shortly.

The transition to 8-dimensional algebras:

When we move into 8-dimensions, counting the relative numbers of 2-dimensional sub-algebras is insufficient to distinguish between different 8-dimensional division algebras. We need also to consider whether the algebra is commutative or non-commutative.

There are seven 4-dimensional sub-algebras within each of the 8-dimensional $C_2 \times C_2 \times C_2$ division algebras. These 4-dimensional sub-algebras are $\{A_1, A_2, A_3, \mathbb{H}\}$. With reference to (12.2) above, those seven 4-dimensional algebras contain respectively the variables:

---

[70] It can be calculated, but guessing is easier.

$$\{a,b,c,d\}, \quad \{a,b,e,f\}, \quad \{a,b,g,h\}, \quad \{a,c,e,g\}$$
$$\{a,c,f,h\}, \quad \{a,d,e,h\}, \quad \{a,d,f,g\}$$

(12.5)

There are only five algebraically non-isomorphic 8-dimensional division algebras. We can distinguish the different algebras by the relative numbers of 2-dimensional division algebras provided we keep track of whether or not the 8-dimensional algebra is commutative or non-commutative; we label the algebras with an appropriate subscript. The 8-dimensional division algebras and their sub-algebras are:

8-dim division algebras	$\sqrt{-1}$	$\sqrt{+1}$	$\mathbb{H}$	A_1	A_2	A_3
$7\sqrt{+1}_{com}$	0	7	0	7	0	0
$3\sqrt{+1}, 4\sqrt{-1}_{com}$	4	3	0	1	6	0
$1\sqrt{+1}, 6\sqrt{-1}_{non-com}$	6	1	4	0	3	0
$3\sqrt{+1}, 4\sqrt{-1}_{non-com}$	4	3	1	0	3	3
$5\sqrt{+1}, 2\sqrt{-1}_{non-com}$	2	5	0	2	1	4

Of course, the A_1 & A_2 algebras are commutative sub-algebras. We have $\left\{1, \vec{e_1}, \vec{e_{23}}, \vec{e_{123}}\right\}$, and this is a commutative Clifford sub-algebra; it is not conventionally included in the set of Clifford algebras. When we rewrite Clifford algebras as division algebras, the commutative sub-algebras are naturally included in the set of $C_2 \times C_2 \times ...$ algebras.

Sub-algebras of the 8-dimensional Clifford algebras:
The 8-dimensional Clifford algebras have the following 2-dimensional and 4-dimensional sub-algebras:

8-dim Clif. algebra	$\sqrt{-1}$	$\sqrt{+1}$	$\mathbb{H}$	A_1	A_2	A_3
$Cl_{0,3}$ $\begin{pmatrix} 1\sqrt{+1} \\ 6\sqrt{-1} \end{pmatrix}$	6	1	$\left\{ \begin{array}{c} 1,\ \vec{e_1} \\ \vec{e_2},\ \vec{e_{12}} \end{array} \right\}$ $\left\{ \begin{array}{c} 1,\ \vec{e_1} \\ \vec{e_3},\ \vec{e_{13}} \end{array} \right\}$ $\left\{ \begin{array}{c} 1,\ \vec{e_2} \\ \vec{e_3},\ \vec{e_{23}} \end{array} \right\}$ $\left\{ \begin{array}{c} 1,\ \vec{e_{12}} \\ \vec{e_{13}},\ \vec{e_{23}} \end{array} \right\}$	0	$\left\{ \begin{array}{c} 1,\ \vec{e_1} \\ \vec{e_{23}},\ \vec{e_{123}} \end{array} \right\}$ $\left\{ \begin{array}{c} 1,\ \vec{e_2} \\ \vec{e_{13}},\ \vec{e_{123}} \end{array} \right\}$ $\left\{ \begin{array}{c} 1,\ \vec{e_3} \\ \vec{e_{12}},\ \vec{e_{123}} \end{array} \right\}$	0
$Cl_{3,0}$ $\begin{pmatrix} 3\sqrt{+1} \\ 4\sqrt{-1} \end{pmatrix}$	4	3	$\left\{ \begin{array}{c} 1,\ \vec{e_{12}} \\ \vec{e_{13}},\ \vec{e_{23}} \end{array} \right\}$	0	$\left\{ \begin{array}{c} 1,\ \vec{e_1} \\ \vec{e_{23}},\ \vec{e_{123}} \end{array} \right\}$ $\left\{ \begin{array}{c} 1,\ \vec{e_2} \\ \vec{e_{13}},\ \vec{e_{123}} \end{array} \right\}$ $\left\{ \begin{array}{c} 1,\ \vec{e_3} \\ \vec{e_{12}},\ \vec{e_{123}} \end{array} \right\}$	$\left\{ \begin{array}{c} 1,\ \vec{e_1} \\ \vec{e_2},\ \vec{e_{12}} \end{array} \right\}$ $\left\{ \begin{array}{c} 1,\ \vec{e_1} \\ \vec{e_3},\ \vec{e_{13}} \end{array} \right\}$ $\left\{ \begin{array}{c} 1,\ \vec{e_2} \\ \vec{e_3},\ \vec{e_{23}} \end{array} \right\}$
$Cl_{1,2}$ $\begin{pmatrix} 3\sqrt{+1} \\ 4\sqrt{-1} \end{pmatrix}$	4	3	$\left\{ \begin{array}{c} 1,\ \vec{e_2} \\ \vec{e_3},\ \vec{e_{23}} \end{array} \right\}$	0	$\left\{ \begin{array}{c} 1,\ \vec{e_1} \\ \vec{e_{23}},\ \vec{e_{123}} \end{array} \right\}$ $\left\{ \begin{array}{c} 1,\ \vec{e_2} \\ \vec{e_{13}},\ \vec{e_{123}} \end{array} \right\}$ $\left\{ \begin{array}{c} 1,\ \vec{e_3} \\ \vec{e_{12}},\ \vec{e_{123}} \end{array} \right\}$	$\left\{ \begin{array}{c} 1,\ \vec{e_1} \\ \vec{e_2},\ \vec{e_{12}} \end{array} \right\}$ $\left\{ \begin{array}{c} 1,\ \vec{e_1} \\ \vec{e_3},\ \vec{e_{13}} \end{array} \right\}$ $\left\{ \begin{array}{c} 1,\ \vec{e_{12}} \\ \vec{e_{13}},\ \vec{e_{23}} \end{array} \right\}$

$Cl_{2,1}$ $\begin{pmatrix}5\sqrt{+1}\\2\sqrt{-1}\end{pmatrix}$	2	5	0	$\left\{\begin{matrix}1, \vec{e_1}\\ \overrightarrow{e_{23}}, \overrightarrow{e_{123}}\end{matrix}\right\}$ $\left\{\begin{matrix}1, \vec{e_2}\\ \overrightarrow{e_{13}}, \overrightarrow{e_{123}}\end{matrix}\right\}$	$\left\{\begin{matrix}1, \vec{e_3}\\ \overrightarrow{e_{12}}, \overrightarrow{e_{123}}\end{matrix}\right\}$	$\left\{\begin{matrix}1, \vec{e_1}\\ \overrightarrow{e_2}, \overrightarrow{e_{12}}\end{matrix}\right\}$ $\left\{\begin{matrix}1, \vec{e_1}\\ \overrightarrow{e_3}, \overrightarrow{e_{13}}\end{matrix}\right\}$ $\left\{\begin{matrix}1, \vec{e_2}\\ \overrightarrow{e_3}, \overrightarrow{e_{23}}\end{matrix}\right\}$ $\left\{\begin{matrix}1, \overrightarrow{e_{12}}\\ \overrightarrow{e_{13}}, \overrightarrow{e_{23}}\end{matrix}\right\}$

We note that the even sub-algebras of $Cl_{3,0}$ & $Cl_{0,3}$ are quaternions and the even sub-algebras of $Cl_{1,2}$ & $Cl_{2,1}$ are, subject to taking the exponential, A_3 algebras.

Taking the equivalent 4-dimensional division algebras, we have the 8-dimensional Clifford algebras as:

8-dim Clifford algebras	$\sqrt{-1}$	$\sqrt{+1}$	$\mathbb{H}$	A_1	A_2	A_3
$Cl_{0,3}\left(1\sqrt{+1}, 6\sqrt{-1}\right)$	6	1	4	0	3	0
$Cl_{3,0}\left(3\sqrt{+1}, 4\sqrt{-1}\right)$	4	3	1	0	3	3
$Cl_{1,2}\left(3\sqrt{+1}, 4\sqrt{-1}\right)$	4	3	1	0	3	3
$Cl_{2,1}\left(5\sqrt{+1}, 2\sqrt{-1}\right)$	2	5	0	2	1	4

We see the $Cl_{3,0}$ & $Cl_{1,2}$ Clifford algebras are algebraically isomorphic. (Only if we interpret multi-vectors to be different from vectors are these two algebras different.) They do have different even sub-algebras, but this is merely whether we call an element a vector or a bi-vector. We have:

$$Cl_{3,0} \sim Cl_{1,2} \sim \left\{1, 3\sqrt{+1}, 4\sqrt{-1}\right\}$$

$$Cl_{0,3} \sim \left\{1, \sqrt{+1}, 6\sqrt{-1}\right\} \qquad (12.6)$$

$$Cl_{2,1} \sim \left\{1, 5\sqrt{+1}, 2\sqrt{-1}\right\}$$

Ignoring the commutative 8-dimensional division algebras, we have a one-to-one correspondence between the 8-dimensional Clifford algebras and the 8-dimensional division algebras.

The 8-dimensional division algebras are generated by three basis elements. The 8-dimensional Clifford algebras are generated by three basis elements (basis vectors). If the natures of these two sets of basis elements match each other, it is hardly surprising that the remaining parts of the algebraic structures also match each other.

The $5\sqrt{+1}, 2\sqrt{-1}_{non-com}$ division algebra:

There are $7 \times 48 = 336$ 8-dimensional[71] isomorphic division algebras of the form $5\sqrt{+1}, 2\sqrt{-1}_{non-com}$. The difference between them is that they are written in different bases connected by unitary transformations. We choose, at random, one of these 336 algebras, and, having considered the relative numbers of square roots of plus unity and square roots of minus unity in both the chosen algebra and the set of 8-dimensional Clifford algebras, we assert that this division algebra is isomorphic to $Cl_{2,1}$. The matrix form is:

---

[71] The 7 arises from the 7 of 8 permutations $\left\{\pm1, \pm1, \pm1\right\}$ of the three quadratic solutions to the scaling parameter equations that give non-commutative algebras.

$5\sqrt{+1}, 2\sqrt{-1}_{non-com} =$

$$\exp\left(\begin{bmatrix} a & b & c & d & e & f & g & h \\ b & a & d & c & f & e & h & g \\ c & -d & a & -b & g & -h & e & -f \\ -d & c & -b & a & -h & g & -f & e \\ e & f & g & h & a & b & c & d \\ f & e & h & g & b & a & d & c \\ g & -h & e & -f & c & -d & a & -b \\ -h & g & -f & e & -d & c & -b & a \end{bmatrix}\right) \qquad (12.7)$$

Not considering the identity, a, we see that we have five symmetric variables $\{b,c,e,f,g\}$ and two anti-symmetric variables $\{d,h\}$. Within the $Cl_{2,1}$ algebra, we have:

$$\begin{aligned} Cl_{2,1} \qquad &: \quad 1, \; \vec{e_1} = \sqrt{+1}, \; \vec{e_2} = \sqrt{+1}, \; \vec{e_3} = \sqrt{-1}, \\ &\vec{e_{12}} = \sqrt{-1}, \; \vec{e_{13}} = \sqrt{+1}, \; \vec{e_{23}} = \sqrt{+1}, \; \vec{e_{123}} = \sqrt{+1} \end{aligned} \qquad (12.8)$$

We immediately identify the identities with each other. Because $\vec{e_{123}}$ commutes with every other basis element of $Cl_{2,1}$ and the element e of the division algebra also commutes with every other basis element of the division algebra, and, in each algebra, there is only one such commutative element other than the identity, we can identify these two basis elements with each other:

$$\begin{aligned} 1 &\equiv a \\ \vec{e_{123}} &\equiv e \end{aligned} \qquad (12.9)$$

Aside:
The Clifford algebra $Cl_{3,0}$ can be written as 2×2 matrices with complex elements. The basis vectors correspond to the Pauli matrices:

$$\vec{e_1} = \begin{bmatrix} 0 & 1 \\ 1 & 0 \end{bmatrix}, \quad \vec{e_2} = \begin{bmatrix} 0 & -i \\ i & 0 \end{bmatrix}, \quad \vec{e_3} = \begin{bmatrix} 1 & 0 \\ 0 & -1 \end{bmatrix} \qquad (12.10)$$

The product of these three matrices is:

$$\begin{bmatrix} 0 & 1 \\ 1 & 0 \end{bmatrix}\begin{bmatrix} 0 & -i \\ i & 0 \end{bmatrix}\begin{bmatrix} 1 & 0 \\ 0 & -1 \end{bmatrix} = \begin{bmatrix} i & 0 \\ 0 & i \end{bmatrix} \qquad (12.11)$$

This is obviously commutative.

Back to the Clifford algebra:
Identifying the other variables is guesswork. However, we know that the anti-symmetric variables are the square roots of minus unity and so either $\vec{e_3} \equiv d$ & $\vec{e_{12}} \equiv h$ or $\vec{e_3} \equiv h$ & $\vec{e_{12}} \equiv d$. We choose the former. This gives:

$$\begin{bmatrix} 1 & \sim & \sim & \vec{e_3} & \vec{e_{123}} & \sim & \sim & \vec{e_{12}} \end{bmatrix} \qquad (12.12)$$

We know that $\vec{e_{12}}\vec{e_3} = \vec{e_3}\vec{e_{12}} = \vec{e_{123}}$, but when we test this we find the above (12.12) gives the negative of what we need. We therefore presume that, in the basis of this particular algebra, one of the basis elements must be negative. We choose the $\vec{e_{12}}$ to be negative:

$$\begin{bmatrix} 1 & \sim & \sim & \vec{e_3} & \vec{e_{123}} & \sim & \sim & -\vec{e_{12}} \end{bmatrix} \qquad (12.13)$$

These four elements form a commutative 4-dimensional sub-algebra. It is a commutative sub-algebra with two anti-symmetric variables (square roots of minus unity) and one symmetric variable (a square root of plus unity); it is the A_2 sub-algebra.

Based on $\vec{e_1}\vec{e_{23}} = \vec{e_{23}}\vec{e_1} = \vec{e_{123}}$, we see that we have the same relations between the $\{bf = e\}$ variables, and so we guess:

$$\vec{e_1} \equiv b$$

$$\overline{e_{23}} \equiv f \tag{12.14}$$

Considering similar relations, with trial and error, this gives:

$$\begin{bmatrix} 1 & \vec{e_1} & \vec{e_{13}} & \vec{e_3} & \overline{e_{123}} & \overline{e_{23}} & -\vec{e_2} & -\overline{e_{12}} \end{bmatrix} \tag{12.15}$$

This exactly matches the Clifford algebra relations. We can therefore write the Clifford algebra $Cl_{2,1}$ as:

$$\begin{bmatrix}
1 & \vec{e_1} & \vec{e_{13}} & \vec{e_3} & \overline{e_{123}} & \overline{e_{23}} & -\vec{e_2} & -\overline{e_{12}} \\
\vec{e_1} & 1 & \vec{e_3} & \vec{e_{13}} & \overline{e_{23}} & \overline{e_{123}} & -\overline{e_{12}} & -\vec{e_2} \\
\vec{e_{13}} & -\vec{e_3} & 1 & -\vec{e_1} & -\vec{e_2} & \vec{e_{12}} & \overline{e_{123}} & -\overline{e_{23}} \\
-\vec{e_3} & \vec{e_{13}} & -\vec{e_1} & 1 & \vec{e_{12}} & -\vec{e_2} & -\overline{e_{23}} & \overline{e_{123}} \\
\overline{e_{123}} & \overline{e_{23}} & -\vec{e_2} & -\overline{e_{12}} & 1 & \vec{e_1} & \vec{e_{13}} & \vec{e_3} \\
\overline{e_{23}} & \overline{e_{123}} & -\overline{e_{12}} & -\vec{e_2} & \vec{e_1} & 1 & \vec{e_3} & \vec{e_{13}} \\
-\vec{e_2} & \vec{e_{12}} & \overline{e_{123}} & -\overline{e_{23}} & \vec{e_{13}} & -\vec{e_3} & 1 & -\vec{e_1} \\
\overline{e_{12}} & -\vec{e_2} & -\overline{e_{23}} & \overline{e_{123}} & -\vec{e_3} & \vec{e_{13}} & -\vec{e_1} & 1
\end{bmatrix} \tag{12.16}$$

Of course, the general Clifford algebra $Cl_{2,1}$ is given by attaching a real variable to each basis element. We must also forget the Clifford algebra relations between the individual basis elements as these are now encoded in the matrix form. The centre of $Cl_{2,1}$ is $\{1, \overline{e_{123}}\}$. The centre of the above $5\sqrt{+1}, 2\sqrt{-1}_{non-com}$ division algebra is $\{1, e\}$ which matches, of course. Taking the exponential of (12.16) will rid it of the singular matrices and thereby change this Clifford algebra into a division algebra, also of course.

The essence of the immediately above is not the calculational procedure but the fact that the Clifford algebra can be written as a division algebra in Cartesian form.

The Clifford product $Cl_{2,1}$:

The Clifford product within $Cl_{2,1}$ is:

$$\left(a\vec{e_1}+b\vec{e_2}+c\vec{e_3}\right)\left(x\vec{e_1}+y\vec{e_2}+z\vec{e_3}\right)=\left(ax+by-cz\right)+\left(ay-bx\right)\vec{e_1}\vec{e_2}$$
$$+\left(az-cx\right)\vec{e_1}\vec{e_3}+\left(bz-cy\right)\vec{e_2}\vec{e_3}$$
$$\equiv \vec{A}\bullet\vec{X}+\vec{A}\times\vec{X}$$

(12.17)

We have:

$$\begin{bmatrix} 0 & a\vec{e_1} & 0 & c\vec{e_3} & 0 & 0 & -b\vec{e_2} & 0 \\ a\vec{e_1} & 0 & c\vec{e_3} & 0 & 0 & 0 & 0 & -b\vec{e_2} \\ 0 & -c\vec{e_3} & 0 & -a\vec{e_1} & -b\vec{e_2} & 0 & 0 & 0 \\ -c\vec{e_3} & 0 & -a\vec{e_1} & 0 & 0 & -b\vec{e_2} & 0 & 0 \\ 0 & 0 & -b\vec{e_2} & 0 & 0 & a\vec{e_1} & 0 & c\vec{e_3} \\ 0 & 0 & 0 & -b\vec{e_2} & a\vec{e_1} & 0 & c\vec{e_3} & 0 \\ -b\vec{e_2} & 0 & 0 & 0 & 0 & -c\vec{e_3} & 0 & -a\vec{e_1} \\ 0 & -b\vec{e_2} & 0 & 0 & -c\vec{e_3} & 0 & -a\vec{e_1} & 0 \end{bmatrix}$$

(12.18)

Taking the product of this matrix (12.18) with a similar one containing the variables $\{x, y, z\}$ in the order of the above Clifford product (12.17) leads to the Clifford product given above.

The basis vectors generate the algebra:

We note that the three basis vectors correspond to variables that do not form a 4-dimensional sub-algebra. This is another way of saying that they generate the entire algebra.

Aside:

The conventional norm of the $Cl_{3,0}$ algebra is[72]:

$$u = u_0 + u_1 \vec{e_1} + u_2 \vec{e_2} + u_3 \vec{e_3} + u_{12} \vec{e_{12}} + u_{13} \vec{e_{13}} + u_{23} \vec{e_{23}} + u_{123} \vec{e_{123}}$$

$$|u|^2 = (u_0)^2 + (u_1)^2 + (u_2)^2 + (u_3)^2 + (u_{12})^2 + (u_{13})^2 + (u_{23})^2 + (u_{123})^2$$

$$(12.19)$$

However, this norm does not arise naturally; instead it is invented and placed upon the Cl_3 algebra by humankind. This is not the norm of any division algebra, and so we cannot form a division algebra equivalent to this Clifford algebra if we insist upon this norm. For that reason, we reject this man-made norm and will prefer the norm that arises naturally from the appropriate division algebra. That norm is the determinant of the matrix form.

Why would a Clifford algebraist choose the above, (12.19), all positive signature quadratic form to be the norm of an 8-dimensional algebra? It is based upon the $\mathbb{R}^n$ view of space that allows us to form a sub-space by 'ripping off' a dimension or two by simply setting a variable or two to zero. Starting with the 2-dimensional all positive signature quadratic form distance function of the complex numbers, extra dimensions are added by adding another variable in such a way that setting any variable to zero will produce a *bona fide* sub-space. The only way to do this is to add another squared variable to the distance function. In the division algebra spaces, we cannot just 'rip off' a dimension to form a *bona fide* sub-space but are restricted by the sub-group structure of the finite group that underlies the division algebra.

The $3\sqrt{+1}, 4\sqrt{-1}_{non-com}$ division algebra:

There are $7 \times 64 = 448$ $3\sqrt{+1}, 4\sqrt{-1}_{non-com}$ division algebras. These are algebraically isomorphic and are simply the same algebra written in 448 different bases. A $3\sqrt{+1}, 4\sqrt{-1}_{non-com}$ division algebra matrix is:

---

[72] Lounesto page 57

$$3\sqrt{+1}, 4\sqrt{-1}_{non-com} =$$

$$\exp\left(\begin{bmatrix} a & b & c & d & e & f & g & h \\ b & a & d & c & f & e & h & g \\ c & -d & a & -b & g & -h & e & -f \\ -d & c & -b & a & -h & g & -f & e \\ -e & -f & -g & -h & a & b & c & d \\ -f & -e & -h & -g & b & a & d & c \\ -g & h & -e & f & c & -d & a & -b \\ h & -g & f & -e & -d & c & -b & a \end{bmatrix}\right) \qquad (12.20)$$

With this matrix, (12.20), ignoring the exponential operator, the Clifford algebra $Cl_{3,0}$ has the corresponding matrix form:

$$\begin{bmatrix} 1 & \vec{e}_1 & -\vec{e}_2 & -\vec{e}_{12} & \vec{e}_{123} & \vec{e}_{23} & \vec{e}_{13} & \vec{e}_3 \end{bmatrix} \\ \equiv \begin{bmatrix} a & b & c & d & e & f & g & h \end{bmatrix} \qquad (12.21)$$

With this matrix, (12.20), the Clifford algebra $Cl_{1,2}$ has the corresponding matrix form:

$$\begin{bmatrix} 1 & \vec{e}_1 & \vec{e}_{12} & \vec{e}_2 & \vec{e}_{123} & \vec{e}_{23} & \vec{e}_3 & \vec{e}_{13} \end{bmatrix} \\ \equiv \begin{bmatrix} a & b & c & d & e & f & g & h \end{bmatrix} \qquad (12.22)$$

The $1\sqrt{+1}, 6\sqrt{-1}_{non-com}$ division algebra:

There are $7 \times 16 = 112$ $1\sqrt{+1}, 6\sqrt{-1}_{non-com}$ division algebras. These are algebraically isomorphic and are simply the same algebra written in 112 different bases. A $1\sqrt{+1}, 6\sqrt{-1}_{non-com}$ division algebra matrix is:

$1\sqrt{+1}, 6\sqrt{-1}_{non-com} =$

$$\exp \left(\begin{bmatrix} a & b & c & d & e & f & g & h \\ -b & a & -d & c & -f & e & -h & g \\ -c & d & a & -b & -g & h & e & -f \\ -d & -c & b & a & -h & -g & f & e \\ e & f & g & h & a & b & c & d \\ -f & e & -h & g & -b & a & -d & c \\ -g & h & e & -f & -c & d & a & -b \\ -h & -g & f & e & -d & -c & b & a \end{bmatrix} \right) \qquad (12.23)$$

Ignoring the exponential operator, this matrix, (12.23), is the Clifford algebra $Cl_{0,3}$ with the corresponding matrix form:

$$\begin{bmatrix} 1 & \overrightarrow{e_1} & \overrightarrow{e_3} & \overrightarrow{e_{13}} & -\overrightarrow{e_{123}} & \overrightarrow{e_{23}} & \overrightarrow{e_{12}} & -\overrightarrow{e_2} \end{bmatrix}$$
$$\equiv \begin{bmatrix} a & b & c & d & e & f & g & h \end{bmatrix} \qquad (12.24)$$

Conclusions:

The difference between two algebraically isomorphic Clifford algebras is in the interpretation and not in the mathematics. It is the interpretation that declares vectors to be different from bi-vectors and other multi-vectors; mathematically, both vectors and multi-vectors are equal basis elements of an algebra.

A Clifford algebra fails to be a division algebra only because not every element of it has a multiplicative inverse and/or it contains zero divisors. Ignoring the exponential operator, a matrix such as (12.23) completely expresses the Clifford algebra including the failure to have a multiplicative inverse for every element of the algebra[73]. Taking the exponential of the matrix rids us of the singular matrices and thus converts the Clifford algebra into a non-commutative division algebra.

---

[73] The matrix referred to can be singular.

There is a one-to-one correspondence between the 8-dimensional Clifford algebras and the non-commutative 8-dimensional $C_2 \times C_2 \times C_2$ division algebras. There are two commutative $C_2 \times C_2 \times C_2$ division algebras that are nothing to do with the Clifford algebras.

There is correspondence between the Clifford algebras and the division algebras because they are both generated by the same number of basis elements $\{\vec{e_1}, \vec{e_2}, \vec{e_3}, ...\}$ which are anti-commutative square roots of plus unity or minus unity[74].

It's not really Clifford algebra, but...:
If we take the relations:

$$\vec{e_1} = \sqrt{+1}, \quad \vec{e_2} = \sqrt{+1}, \quad \vec{e_3} = \sqrt{+1}$$
$$\vec{e_1}\vec{e_2} = \vec{e_2}\vec{e_1}, \quad \vec{e_1}\vec{e_3} = \vec{e_3}\vec{e_1}, \quad \vec{e_2}\vec{e_3} = \vec{e_3}\vec{e_2} \qquad (12.25)$$

in which the vectors are commutative, we get the commutative $7\sqrt{+1}_{com}$ division algebra. If we take the relations:

$$\vec{e_1} = \sqrt{-1}, \quad \vec{e_2} = \sqrt{-1}, \quad \vec{e_3} = \sqrt{-1}$$
$$\vec{e_1}\vec{e_2} = \vec{e_2}\vec{e_1}, \quad \vec{e_1}\vec{e_3} = \vec{e_3}\vec{e_1}, \quad \vec{e_2}\vec{e_3} = \vec{e_3}\vec{e_2} \qquad (12.26)$$

in which the vectors are commutative, we get the commutative $3\sqrt{+1}, 4\sqrt{-1}_{com}$ division algebra.

---

[74] There are other 8-dimensional division algebras; the C_8 division algebras have basis elements that are the 8th roots of plus unity or minus unity and the $C_2 \times C_4$ division algebras have basis elements that are either the square roots of plus or minus unity or are the 4th roots of plus or minus unity.

The story so far:

We are saying that the whole of Clifford algebra is contained within the higher dimensional division algebras (the higher dimensional complex numbers). There are more division algebras than there are Clifford algebras; this is not only the commutative division algebras and the anti-algebras as will be clearer later. We are saying that the whole of the Clifford algebras is contained within the symmetric and anti-symmetric matrices that form the $C_2 \times C_2 \times ...$ division algebras.

We note that the symmetric matrices have real eigenvalues and orthogonal eigenvectors and so can perhaps be used to replace the hermitian matrices we usually find in quantum mechanics. We will shortly see that the division algebra spaces are spinor spaces, and that the $C_2 \times C_2 \times ...$ division algebras are spinors which double cover the $SO(p,q)$ groups.

Chapter 13

An 8-dimensional Rotation Matrix

We seek to calculate the rotation matrix (polar form) of the 8-dimensional algebra[75]:

$$DQ = \exp\left(\begin{bmatrix} a & b & c & d & e & f & g & h \\ -b & a & -d & c & -f & e & -h & g \\ -c & d & a & -b & -g & h & e & -f \\ -d & -c & b & a & -h & -g & f & e \\ e & f & g & h & a & b & c & d \\ -f & e & -h & g & -b & a & -d & c \\ -g & h & e & -f & -c & d & a & -b \\ -h & -g & f & e & -d & -c & b & a \end{bmatrix}\right) \qquad (13.1)$$

This is a $Cl_{0,3}$ $\{1,\sqrt{+1},6\sqrt{-1}\}$ algebra. The mathematics computer program Maple 17 seems unable to calculate this polar form all at once using the function *MatrixExponential*(); perhaps Maple just does not like me. However, because:

$$e^{A+B} = e^A e^B \qquad if \qquad AB = BA \qquad (13.2)$$

we can take out the commutative variables $\{a,e\}$ and calculate them separately before multiplying them into the non-commutative rotation matrix, DQ^{NC}. Even with $a = e = 0$, Maple does not always accomplish the calculation, but sometimes it does accomplish the calculation. Maple can be moody, but she loves me really.

The Maple results with $a = e = 0$ are[76]:

---

[75] The name DQ is an abbreviation of Double Quaternion.
[76] We give the top row only for presentational ease.

$$
DQ^{NC}_{[1,1]} = \frac{1}{4} \left(
\frac{\left(e^{\sqrt{-(b+f)^2-(c+g)^2-(d+h)^2}}\right)^2 e^{\sqrt{-(b-f)^2-(c-g)^2-(d-h)^2}}}{e^{\sqrt{-(b+f)^2-(c+g)^2-(d+h)^2}} e^{\sqrt{-(b-f)^2-(c-g)^2-(d-h)^2}}}
\right.
$$

$$
+ \frac{e^{\sqrt{-(b+f)^2-(c+g)^2-(d+h)^2}} \left(e^{\sqrt{-(b-f)^2-(c-g)^2-(d-h)^2}}\right)^2}{e^{\sqrt{-(b+f)^2-(c+g)^2-(d+h)^2}} e^{\sqrt{-(b-f)^2-(c-g)^2-(d-h)^2}}}
$$

$$
+ \frac{e^{\sqrt{-(b-f)^2-(c-g)^2-(d-h)^2}}}{e^{\sqrt{-(b+f)^2-(c+g)^2-(d+h)^2}} e^{\sqrt{-(b-f)^2-(c-g)^2-(d-h)^2}}}
$$

$$
\left. + \frac{e^{\sqrt{-(b+f)^2-(c+g)^2-(d+h)^2}}}{e^{\sqrt{-(b+f)^2-(c+g)^2-(d+h)^2}} e^{\sqrt{-(b-f)^2-(c-g)^2-(d-h)^2}}}
\right)
$$

$$
= \frac{1}{4}\left(e^{\sqrt{-(b+f)^2-(c+g)^2-(d+h)^2}} + e^{\sqrt{-(b-f)^2-(c-g)^2-(d-h)^2}} + \frac{1}{e^{\sqrt{-(b+f)^2-(c+g)^2-(d+h)^2}}} + \frac{1}{e^{\sqrt{-(b-f)^2-(c-g)^2-(d-h)^2}}} \right)
$$

$$
= \frac{1}{2}\left(
\begin{array}{c}
\cosh\left(\sqrt{-(b+f)^2-(c+g)^2-(d+h)^2}\right) \\
+\cosh\left(\sqrt{-(b-f)^2-(c-g)^2-(d-h)^2}\right)
\end{array}
\right)
$$

$$
= \frac{1}{2}\left(
\begin{array}{c}
\cosh\left(i\sqrt{(b+f)^2+(c+g)^2+(d+h)^2}\right) \\
+\cosh\left(i\sqrt{(b-f)^2+(c-g)^2+(d-h)^2}\right)
\end{array}
\right) \tag{13.3}
$$

Which is:

$$
DQ^{NC}_{[1,1]} = \frac{1}{2}\left(
\begin{array}{c}
\cos\left(\sqrt{(b+f)^2+(c+g)^2+(d+h)^2}\right) \\
+\cos\left(\sqrt{(b-f)^2+(c-g)^2+(d-h)^2}\right)
\end{array}
\right) \tag{13.4}
$$

Of course, we must multiply the rotation matrix by the exponentials of the $\{a,e\}$ variables to get the complete polar form of the algebra. We will do this later.

The other results are:

$$DQ^{NC}_{[1,2]} = \frac{1}{2}\left(\frac{(b+f)\sin\left(\sqrt{(b+f)^2+(c+g)^2+(d+h)^2}\right)}{\sqrt{(b+f)^2+(c+g)^2+(d+h)^2}} + \frac{(b-f)\sin\left(\sqrt{(b-f)^2+(c-g)^2+(d-h)^2}\right)}{\sqrt{(b-f)^2+(c-g)^2+(d-h)^2}} \right) \tag{13.5}$$

$$DQ^{NC}_{[1,3]} = \frac{1}{2}\left(\frac{(c+g)\sin\left(\sqrt{(b+f)^2+(c+g)^2+(d+h)^2}\right)}{\sqrt{(b+f)^2+(c+g)^2+(d+h)^2}} + \frac{(c-g)\sin\left(\sqrt{(b-f)^2+(c-g)^2+(d-h)^2}\right)}{\sqrt{(b-f)^2+(c-g)^2+(d-h)^2}} \right) \tag{13.6}$$

$$DQ^{NC}_{[1,4]} = \frac{1}{2}\left(\frac{(d+h)\sin\left(\sqrt{(b+f)^2+(c+g)^2+(d+h)^2}\right)}{\sqrt{(b+f)^2+(c+g)^2+(d+h)^2}} + \frac{(d-h)\sin\left(\sqrt{(b-f)^2+(c-g)^2+(d-h)^2}\right)}{\sqrt{(b-f)^2+(c-g)^2+(d-h)^2}} \right) \tag{13.7}$$

$$DQ^{NC}_{[1,5]} = \frac{1}{2}\left(\cos\left(\sqrt{(b+f)^2+(c+g)^2+(d+h)^2}\right) - \cos\left(\sqrt{(b-f)^2+(c-g)^2+(d-h)^2}\right) \right) \tag{13.8}$$

$$DQ^{NC}_{[1,6]} = \frac{1}{2}\left(\frac{(b+f)\sin\left(\sqrt{(b+f)^2+(c+g)^2+(d+h)^2}\right)}{\sqrt{(b+f)^2+(c+g)^2+(d+h)^2}} - \frac{(b-f)\sin\left(\sqrt{(b-f)^2+(c-g)^2+(d-h)^2}\right)}{\sqrt{(b-f)^2+(c-g)^2+(d-h)^2}} \right) \tag{13.9}$$

$$DQ^{NC}_{[1,7]} = \frac{1}{2} \left(\begin{array}{c} \dfrac{(c+g)\sin\left(\sqrt{(b+f)^2+(c+g)^2+(d+h)^2}\right)}{\sqrt{(b+f)^2+(c+g)^2+(d+h)^2}} \\[2em] -\dfrac{(c-g)\sin\left(\sqrt{(b-f)^2+(c-g)^2+(d-h)^2}\right)}{\sqrt{(b-f)^2+(c-g)^2+(d-h)^2}} \end{array} \right) \tag{13.10}$$

$$DQ^{NC}_{[1,8]} = \frac{1}{2} \left(\begin{array}{c} \dfrac{(d+h)\sin\left(\sqrt{(b+f)^2+(c+g)^2+(d+h)^2}\right)}{\sqrt{(b+f)^2+(c+g)^2+(d+h)^2}} \\[2em] -\dfrac{(d-h)\sin\left(\sqrt{(b-f)^2+(c-g)^2+(d-h)^2}\right)}{\sqrt{(b-f)^2+(c-g)^2+(d-h)^2}} \end{array} \right) \tag{13.11}$$

Note the single sign difference of the second terms in each of the last four expressions compared with the first four expressions.

If we were to set all the imaginary variables to zero except one of the $\{b,c,d,f.g.h\}$, we would have a double cover of six 2-dimensional rotations within this algebra.

We now have to multiply this matrix by the exponentials of the commutative $\{a,e\}$ variables to get the polar form of the algebra. The exponential of the a variable provides nothing other than the real number which is the radial component of the algebra:

$$\exp\left(\begin{bmatrix} a & \dots & 0 \\ \dots & \dots & \dots \\ 0 & \dots & a \end{bmatrix}\right) = \begin{bmatrix} e^a & \dots & 0 \\ \dots & \dots & \dots \\ 0 & \dots & e^a \end{bmatrix} \tag{13.12}$$

The exponential of the e variable is:

$$\exp([e]) = \begin{bmatrix} \cosh(e) & 0 & 0 & 0 & \sinh(e) & 0 & 0 & 0 \\ 0 & \sim & 0 & 0 & 0 & \sim & 0 & 0 \\ 0 & 0 & \sim & 0 & 0 & 0 & \sim & 0 \\ 0 & 0 & 0 & \cosh(e) & 0 & 0 & 0 & \sinh(e) \\ \sinh(e) & 0 & 0 & 0 & \cosh(e) & 0 & 0 & 0 \\ 0 & \sim & 0 & 0 & 0 & \sim & 0 & 0 \\ 0 & 0 & \sim & 0 & 0 & 0 & \sim & 0 \\ 0 & 0 & 0 & \sinh(e) & 0 & 0 & 0 & \cosh(e) \end{bmatrix}$$

$$(13.13)$$

Multiplying the e exponential matrix into the non-commutative matrix gives the complete 8-dimensional rotation matrix:

$$DQ_{[1,1]}^{ROT} = \cosh(e) DQ_{[1,1]}^{NC} + \sinh(e) DQ_{[1,5]}^{NC}$$
$$DQ_{[1,2]}^{ROT} = \cosh(e) DQ_{[1,2]}^{NC} + \sinh(e) DQ_{[1,6]}^{NC}$$
$$DQ_{[1,3]}^{ROT} = \cosh(e) DQ_{[1,3]}^{NC} + \sinh(e) DQ_{[1,7]}^{NC}$$
$$DQ_{[1,4]}^{ROT} = \cosh(e) DQ_{[1,4]}^{NC} + \sinh(e) DQ_{[1,8]}^{NC}$$

$$(13.14)$$

And:

$$DQ_{[1,5]}^{ROT} = \cosh(e) DQ_{[1,5]}^{NC} + \sinh(e) DQ_{[1,1]}^{NC}$$
$$DQ_{[1,6]}^{ROT} = \cosh(e) DQ_{[1,6]}^{NC} + \sinh(e) DQ_{[1,2]}^{NC}$$
$$DQ_{[1,7]}^{ROT} = \cosh(e) DQ_{[1,7]}^{NC} + \sinh(e) DQ_{[1,3]}^{NC}$$
$$DQ_{[1,8]}^{ROT} = \cosh(e) DQ_{[1,8]}^{NC} + \sinh(e) DQ_{[1,4]}^{NC}$$

$$(13.15)$$

This is an 8-dimensional rotation (rotation in a spinor space with eight independent components equivalent to four $\mathbb{C}$ complex numbers[77]). It is the case in 8-dimensions that, as in general, a rotation through

---

[77] Do not think that this is a Dirac spinor. The gamma matrices are the one basis vector and three basis bivectors of a 16-dimensional Clifford algebra.

$\{b,c,d,e,f,g,h\}$ which is followed by a rotation through $\{-b,-c,-d,-e,-f,-g,-h\}$ gives the identity.

Eigenvalues:

Eigenvalues of the Cartesian form of the matrix, (13.1), are:

$$
\begin{bmatrix}
a-e+\hat{i}\sqrt{(b-f)^2+(c-g)^2+(d-h)^2} \\
a-e-\hat{i}\sqrt{(b-f)^2+(c-g)^2+(d-h)^2} \\
a+e+\hat{i}\sqrt{(b+f)^2+(c+g)^2+(d+h)^2} \\
a+e-\hat{i}\sqrt{(b+f)^2+(c+g)^2+(d+h)^2} \\
a-e+\hat{i}\sqrt{(b-f)^2+(c-g)^2+(d-h)^2} \\
a-e-\hat{i}\sqrt{(b-f)^2+(c-g)^2+(d-h)^2} \\
a+e+\hat{i}\sqrt{(b+f)^2+(c+g)^2+(d+h)^2} \\
a+e-\hat{i}\sqrt{(b+f)^2+(c+g)^2+(d+h)^2}
\end{bmatrix}
\qquad (13.16)
$$

$$\hat{i}=\sqrt{-1}$$

Note that we have put a little hat over the $\hat{i}=\sqrt{-1}$ to avoid confusing it with the many variables, none of which is denoted by i or by o.

We notice that the e variable 'thinks' it is real variable. Within the eigenvalues, its function is exactly the same as the real variable, a. Similarly, within the eigenvalues, the pair of variables $(b+f)$ 'think' they are one of the 4-dimensional imaginary variables. It seems as if 8-dimensional space 'thinks' it is 4-dimensional space.

Comments:

The DQ, (13.1), algebra is of the form:

$$\begin{bmatrix} \mathbb{H}_1 & \mathbb{H}_2 \\ \mathbb{H}_2 & \mathbb{H}_1 \end{bmatrix} \tag{13.17}$$

If $e = 0$, the 8-dimensional trigonometric functions (13.14) & (13.15) lose their second term because $\sinh(0) = 0$ and they become the purely non-commutative, DQ^{NC}, (13.4) to (13.11) trigonometric functions because $\cosh(0) = 1$. If $e = f = g = h = 0$, these functions become the quaternion trigonometric functions:

$$DQ_{[1,1]} = \frac{1}{2}\left(\begin{array}{c} \cos\left(\sqrt{b^2 + c^2 + d^2}\right) \\ + \cos\left(\sqrt{b^2 + c^2 + d^2}\right) \end{array} \right) = \cos\left(\sqrt{b^2 + c^2 + d^2}\right) \tag{13.18}$$

$$DQ_{[1,2]} = \frac{1}{2}\left(\begin{array}{c} \dfrac{b\sin\left(\sqrt{b^2 + c^2 + d^2}\right)}{\sqrt{b^2 + c^2 + d^2}} \\ + \dfrac{b\sin\left(\sqrt{b^2 + c^2 + d^2}\right)}{\sqrt{b^2 + c^2 + d^2}} \end{array} \right) = \frac{b\sin\left(\sqrt{b^2 + c^2 + d^2}\right)}{\sqrt{b^2 + c^2 + d^2}} \tag{13.19}$$

$$DQ_{[1,5]} = \frac{1}{2}\left(\begin{array}{c} \cos\left(\sqrt{b^2 + c^2 + d^2}\right) \\ - \cos\left(\sqrt{b^2 + c^2 + d^2}\right) \end{array} \right) = 0 \tag{13.20}$$

$$DQ_{[1,6]} = \frac{1}{2}\left(\begin{array}{c} \dfrac{b\sin\left(\sqrt{b^2 + c^2 + d^2}\right)}{\sqrt{b^2 + c^2 + d^2}} \\ - \dfrac{b\sin\left(\sqrt{b^2 + c^2 + d^2}\right)}{\sqrt{b^2 + c^2 + d^2}} \end{array} \right) = 0 \tag{13.21}$$

This is a general phenomenon over all higher dimensions. The top left-hand 4×4 block of any $C_2 \times C_2 \times ...$ division algebra is a 4-dimensional sub-algebra of the higher dimensional algebra. The only such sub-groups are the 4-dimensional $C_2 \times C_2$ algebras $\{\mathbb{H}, A_3, A_2, A_1\}$. We see that all higher dimensional trigonometric functions of the $C_2 \times C_2 \times ...$ algebras must reduce to one set of the 4-dimensional $C_2 \times C_2$ trigonometric functions when all variables other than the first four are zero.

There are other 4-dimensional sub-algebras within this 8-dimensional algebra, (13.1), (and all other 8-dimensional algebras); an example is the variables $\{a,c,f,h\}$. In this case, the trigonometric functions reduce as above (13.18) to (13.21):

$$DQ^{NC}_{[1,1]} = \frac{1}{2}\left(\begin{array}{c} \cos\left(\sqrt{f^2+c^2+h^2}\right) \\ +\cos\left(\sqrt{f^2+c^2+h^2}\right) \end{array} \right) = \cos\left(\sqrt{f^2+c^2+h^2}\right) \qquad (13.22)$$

The 4-dimensional sub-algebras are:

$$\begin{array}{ll} \{a,b,c,d\} & \{a,b,e,f\} \\ \{a,b,g,h\} & \{a,c,e,g\} \\ \{a,c,f,h\} & \{a,d,e,h\} \\ \{a,d,f,g\} & \end{array} \qquad (13.23)$$

The algebras in the first column have $e = 0$ and their trigonometric functions reduce to the appropriate 4-dimensional trigonometric functions as shown above, (13.18) to (13.21).

When $e \neq 0$, we have a commutative sub-algebra which is either a A_1 or a A_2 sub-algebra. In the case of this $Cl_{0,3}$ algebra, we have a A_2 sub-algebra $\{1, \vec{e_1}, \vec{e_{23}}, \vec{e_{123}}\}$. Not surprisingly, the trigonometric functions of the A_2 algebra are not the same as the quaternion trigonometric functions.

The 8-dimensional trigonometric functions are of the form (13.14) & (13.15). For the $\{a,b,e,f\}$ sub-algebra, we have:

$$DQ_{[1,1]}^{ROT} = \cosh(e)\frac{1}{2}\begin{pmatrix} \cos(b+f) \\ +\cos(b-f) \end{pmatrix} + \sinh(e)\frac{1}{2}\begin{pmatrix} \cos(b+f) \\ -\cos(b-f) \end{pmatrix}$$

$$= \cosh(e)\left(\cos(b)\cos(f)\right) - \sinh(e)\left(\sin(b)\sin(f)\right)$$

$$(13.24)$$

Similarly, the 4-dimensional $C_2 \times C_2$ division algebras must accommodate the trigonometric functions of all the 2-dimensional sub-algebras; putting $e = f = 0$ into (13.24) gives $\cos(b)$.

The essence of all this is that the 8-dimensional trigonometric functions of the 8-dimensional $C_2 \times C_2 \times C_2$ division algebras must accommodate the trigonometric functions of all the 4-dimensional and all the 2-dimensional sub-algebras of the particular 8-dimensional algebra. The different possible types of sub-algebras have different trigonometric functions. This severely constrains the possible form of the 8-dimensional trigonometric functions. Similarly, the form of the 16-dimensional trigonometric functions will be constrained by the need to accommodate the trigonometric functions of the lesser dimensional sub-algebras.

The pairing of variables:
The determinant of this algebra, (13.1), is:

$$\det(DQ) = PROD\begin{pmatrix} \left((a+e)^2 + (b+f)^2 + (c+g)^2 + (d+h)^2\right)^2 \\ \left((a-e)^2 + (b-f)^2 + (c-g)^2 + (d-h)^2\right)^2 \end{pmatrix}$$

$$(13.25)$$

In this determinant, which is the distance function of the 8-dimensional space, we again see the pairing of variables that we have previously seen in the trigonometric functions, (13.4) to (13.11), and the eigenvalues, (13.16).

Such pairing of variables is not unique to the $C_2 \times C_2 \times ...$ division algebras. The finite group C_4 gives rise to eight 4-dimensional algebras. A typical trigonometric function from the C_4 algebras is[78]:

$$v A = \frac{1}{2}\left(e^c \cosh(b+d) + e^{-c} \cos(b-d)\right) \qquad (13.26)$$

If the higher dimensional algebras are going to accommodate the 4-dimensional sub-algebras, this pairing of variables is essential, and so we might expect a similar pairing of the 'pairs of variables' in the 16-dimensional algebras and so on into higher dimensions. We consider the 16-dimensional division algebra of the form:

$$\{1, 3\sqrt{+1}, 12\sqrt{-1}\} = \begin{bmatrix} \mathbb{H}_1 & \mathbb{H}_2 & \mathbb{H}_3 & \mathbb{H}_4 \\ \mathbb{H}_2 & \mathbb{H}_1 & \mathbb{H}_4 & \mathbb{H}_3 \\ \mathbb{H}_3 & \mathbb{H}_4 & \mathbb{H}_1 & \mathbb{H}_2 \\ \mathbb{H}_4 & \mathbb{H}_3 & \mathbb{H}_2 & \mathbb{H}_1 \end{bmatrix} \qquad (13.27)$$

This is not a conventional Clifford algebra, but it is one of the Clifford algebras which are missed in the traditional derivation of Clifford algebras – see (15.3) later. The determinant of this algebra, (13.27), is:

$$\det\left(\left\{\begin{bmatrix} 1 \\ 3\sqrt{+1} \\ 12\sqrt{-1} \end{bmatrix}\right\}\right) = PROD \begin{pmatrix} \left(\begin{array}{c}(a-e-j+n)^2 + (b-f-k+p)^2 \\ +(c-g-l+q)^2 + (d-h-m+r)^2\end{array}\right)^2 \\ \left(\begin{array}{c}(a+e+j+n)^2 + (b+f+k+p)^2 \\ +(c+g+l+q)^2 + (d+h+m+r)^2\end{array}\right)^2 \\ \left(\begin{array}{c}(a+e-j-n)^2 + (b+f-k-p)^2 \\ +(c+g-l-q)^2 + (d+h-m-r)^2\end{array}\right)^2 \\ \left(\begin{array}{c}(a-e+j-n)^2 + (b-f+k-p)^2 \\ +(c-g+l-q)^2 + (d-h+m-r)^2\end{array}\right)^2 \end{pmatrix} \qquad (13.28)$$

---

[78] Dennis Morris: Complex Numbers The Higher Dimensional Forms: ISBN: 978-1508877499. Page 145

We see the pairing of 'pairs of variables' within the determinant. Hardly surprisingly, we see this pairing of 'pairs of variables' in the eigenvalues[79] which are of the form:

$$a+e+j+n+\hat{i}\sqrt{\begin{array}{c}(b+f+k+p)^2\\+(c+g+l+q)^2\\+(d+h+m+r)^2\end{array}}$$

(13.29)

$$\hat{i}=\sqrt{-1}$$

Note that we have again put a little hat over the $i=\sqrt{-1}$ to avoid confusing it with the many variables, none of which is denoted by i or o. It is as if the higher dimensional spaces are 'packed together' into a 4-dimensional format.

We have chosen to demonstrate this pairing of 'pairs of variables' using the quaternion matrices because it is so clear within these matrices. If we had chosen to use a conventional Clifford algebra such as $Cl_{1,3}=\{1,9\sqrt{+1},6\sqrt{-1}\}$, the pairing would be far from obvious. Maple 17, on my PC, will not calculate the determinant and eigenvalues of such an algebra, and so we offer a partial determinant in which many of the variables have been set to zero to allow Maple 17 to do the calculation. The partial determinant of a $Cl_{1,3}=\{1,9\sqrt{+1},6\sqrt{-1}\}$ algebra is:

$$\det\left(Cl_{1,3}\right)=\left(\begin{array}{c}a^4+b^4+c^4+e^4+n^4+2\left(\begin{array}{c}a^2b^2+a^2c^2-a^2e^2\\-a^2h^2-a^2n^2\end{array}\right)\\+2\left(\begin{array}{c}-b^2c^2-b^2e^2\\-b^2h^2+b^2n^2\end{array}\right)+2\left(-c^2e^2-c^2h^2-c^2n^2\right)\\+2\left(-e^2h^2-e^2n^2\right)+2h^2n^2+8bceh\end{array}\right)^4$$

(13.30)

Within any 16-dimensional $C_2 \times C_2 \times ...$ division algebra, there are fifteen 8-dimensional sub-algebras, thirty-five 4-dimensional sub-algebras, and fifteen 2-dimensional sub-algebras. The 16-dimensional trigonometric functions have a lot of accommodating to do.

Why do we not see 8-dimensional space?:

We have earlier derived the 4-dimensional space-time in which we sit by superimposition of the six A_3 algebras, and we declared that this is why we live in our 4-dimensional space-time. Why can we not superimpose the isomorphic 8-dimensional algebras and have an 8-dimensional space? We do not properly understand an answer to this question, but we speculate that this pairing of variables might in some way reduce such 8-dimensional superimpositions to a 4-dimensional form. We suspect that the quarks are associated with 8-dimensional spaces. Possibly, this pairing of variables is associated with quark confinement.

6-dimensional spinors:

There are two finite groups of order six. These are firstly the commutative cyclic group C_6, which we do not consider because it contains only commutative algebras, and secondly the non-commutative symmetric group S_3 (isomorphic to D_3). A basic algebraic matrix form of the S_3 group is:

$$S_3 = \exp\left(\begin{bmatrix} a & b & c & d & e & f \\ c & a & b & e & f & d \\ b & c & a & f & d & e \\ d & e & f & a & b & c \\ e & f & d & c & a & b \\ f & d & e & c & b & a \end{bmatrix}\right) \tag{13.31}$$

By setting $a = 0$ and taking the matrix exponential, we get the S_3 rotation matrix containing the trigonometric functions of S_3 space. These are not pretty functions; we do not offend the reader's eye by showing the off-diagonal trigonometric functions. The leading S_3 trigonometric function is much simpler. It is:

$$\frac{v_A(S_3)}{3456} =$$

$$\frac{PROD\left(\begin{array}{c}\left(\begin{array}{c}b(b+c-d-e-f)+c(c-d-e-f)\\+d(e+f)+ef\end{array}\right)^2 \\ \left(\begin{array}{c}b(b+c+d+e+f)+c(c+d+e+f)\\+d(e+f)+ef\end{array}\right)^2 \\ \left(4e^{-\frac{(b+c)}{2}}\cosh\left(\frac{\sqrt{\begin{array}{c}-3(b-c)^2\\+4(d(d-e-f)+e(e-f)+f^2)\end{array}}}{2}\right)\\+2e^{(b+c)}\cosh(d+e+f)\right)\end{array}\right)}{PROD\left(\begin{array}{c}\left(\dfrac{2(d+e+f)-3(b+c)}{+\sqrt{-3(b-c)^2+4(d(d-e-f)+e(e-f)+f^2)}}\right)^2 \\ \left(\dfrac{-2(d+e+f)-3(b+c)}{+\sqrt{-3(b-c)^2+4(d(d-e-f)+e(e-f)+f^2)}}\right)^2 \\ \left(\dfrac{-2(d+e+f)+3(b+c)}{+\sqrt{-3(b+c)^2+4(d(d-e-f)+e(e-f)+f^2)}}\right)^2 \\ \left(\dfrac{2(d+e+f)+3(b+c)}{+\sqrt{-3(b-c)^2+4(d(d-e-f)+e(e-f)+f^2)}}\right)^2 \end{array}\right)} \quad (13.32)$$

We note that when $d = e = f = 0$ this becomes[80]:

---

[80] It is quite remarkable how the 3456 cancels.

$$v_A\left(S_3\right)_{d=e=f=0} = \frac{1}{3}\left(e^{(b+c)} + 2e^{-\frac{(b+c)}{2}} \cos\left(\frac{\sqrt{3}}{2}(b-c) \right) \right) \qquad (13.33)$$

This is the leading trigonometric function of the 3×3 C_3 algebra which sits in the top left-hand corner of (13.31) and is the order 3 sub-group of the S_3 algebra. Of course it is; the 6-dimensional trigonometric functions must reduce to the 3-dimensional trigonometric functions when we set the other variables to zero.

We note that if the expression in the square root in the $\cosh(\ )$ function, (13.32), should be negative, then the $\cosh(\ )$ becomes a $\cos(\ )$ function and this does not introduce an imaginary number into the value of the trigonometric function. Hence, (13.32) is a real number. The 3456 comes out of the calculation; quite what it means, if anything, is not understood.

Double covering – 8-dimensional:
Looking at the 8-dimensional trigonometric functions above, (13.14) & (13.15), we see that, if we consider the 8-dimensional 2-dimensional rotations rotations given by setting all variables to zero except a and one other variable then we see a double cover for the 2-dimensional sub-algebras with the variables $\{b,c,d,f,g,h\}$ but no double cover for the e variable 2-dimensional sub-algebra. The nature of the 8-dimensional 2-dimensional $\{a,e\}$ rotation is that of a space-time rotation, a boost and it is not a double cover.

There are no $\mathbb{R}^n$ spaces with seven pairs of axes; in general, $\mathbb{R}^n$ spaces have $\frac{n(n-1)}{2}$ pairs of axes; but wait, we have only 6 double covers of 2-dimensional rotations in the 8-dimensional algebras. There are six pairs of axes in $\mathbb{R}^4$. It seems that the 8-dimensional spaces double cover the 2-dimensional rotations in $\mathbb{R}^4$ spaces like our space-time. If the calculations did not show it, I would think I was cheating!

Double covering – 6-dimensional:

The S_3 algebra, (13.31), has one 3-dimensional sub-algebra, $\{a,b,c\}$, and three 2-dimensional sub-algebras, $\{a,d\}$, $\{a,e\}$, $\{a,f\}$. We seek double covers of the 2-dimensional rotations because we have only 2-dimensional rotations in Riemannian space[81].

Looking at the 6-dimensional trigonometric function, (13.32), when $b=c=0$, this becomes:

$$V_A\left(S_3\right)_{b=c=0} = \frac{\cosh\left(d+e+f\right)+2\cosh\left(\sqrt{\left(d^2-de-df+e^2-ef+f^2\right)}\right)}{6}$$

$$(13.34)$$

When $e=f=0$, this is:

$$V_A\left(S_3\right)_{b=c=0} = \frac{\cosh\left(d\right)+2\cosh\left(\sqrt{d^2}\right)}{6} \qquad (13.35)$$

Because $\cosh(\ )$ is an even function, this is:

$$V_A\left(S_3\right)_{b=c=0} = \frac{\cosh\left(d\right)}{2} \qquad (13.36)$$

We also have:

$$V_B\left(S_3\right)_{b=c=0} = \frac{\cosh\left(d+e+f\right)-\cosh\left(\sqrt{\left(d^2-de-df+e^2-ef+f^2\right)}\right)}{6}$$

$$(13.37)$$

When $e=f=0$, this is:

$$V_B\left(S_3\right)_{b=c=0} = \frac{\cosh\left(d\right)-\cosh\left(\sqrt{\left(d^2\right)}\right)}{6} = 0 \qquad (13.38)$$

---

[81] The question of double covers of 3-dimensional rotations and of all higher dimensional rotations has never been asked to date.

Because $\cosh(\ )$ is an even function, this is zero regardless of which root of $\sqrt{d^2}$ we choose.

We also have:

$$v_C\left(S_3\right)_{b=c=0} = \frac{\cosh\left(d+e+f\right)-\cosh\left(\sqrt{\left(d^2-de-df+e^2-ef+f^2\right)}\right)}{6}$$

$$(13.39)$$

When $e = f = 0$, this is:

$$v_B\left(S_3\right)_{b=c=0} = \frac{\cosh\left(d\right)-\cosh\left(\sqrt{\left(d^2\right)}\right)}{6} = 0 \qquad (13.40)$$

We also have:

$$v_D\left(S_3\right)_{b=c=0} = \frac{1}{6}\sinh\left(d+e+f\right)$$
$$+\frac{\left(2d-e-f\right)\sinh\left(\sqrt{\left(d^2-de-df+e^2-ef+f^2\right)}\right)}{6\sqrt{\left(d^2-de-df+e^2-ef+f^2\right)}}$$

$$(13.41)$$

When $e = f = 0$, this is:

$$v_D\left(S_3\right)_{b=c=0} = \frac{\sinh\left(d\right)+2\sqrt{+1}\sinh\left(\sqrt{d^2}\right)}{6} \qquad (13.42)$$

We also have:

$$v_E\left(S_3\right)_{b=c=0} = \frac{1}{6}\sinh\left(d+e+f\right)$$
$$+\frac{\left(2e-d-f\right)\sinh\left(\sqrt{\left(d^2-de-df+e^2-ef+f^2\right)}\right)}{6\sqrt{\left(d^2-de-df+e^2-ef+f^2\right)}}$$

$$(13.43)$$

When $e = f = 0$, this is:

$$v_E\left(S_3\right)_{b=c=0} = \frac{\sinh\left(d\right) - \sqrt{+1}\,\sinh\left(\sqrt{d^2}\right)}{6} \qquad (13.44)$$

We also have:

$$v_F\left(S_3\right)_{b=c=0} = \frac{1}{6}\sinh\left(d+e+f\right)$$

$$+ \frac{\left(2f-d-e\right)\sinh\left(\sqrt{\left(d^2-de-df+e^2-ef+f^2\right)}\right)}{6\sqrt{\left(d^2-de-df+e^2-ef+f^2\right)}}$$

$$(13.45)$$

When $e = f = 0$, this is:

$$v_F\left(S_3\right)_{b=c=0} = \frac{\sinh\left(d\right) - \sqrt{+1}\,\sinh\left(\sqrt{d^2}\right)}{6} \qquad (13.46)$$

We know that $\{a,d\}$ is a 2-dimensional sub-algebra, and so we know that, when $b = c = e = f = 0$, we should have $v_E\left(S_3\right) = v_F\left(S_3\right) = 0$. We can achieve this if we take both square roots to be positive or both square roots to be negative but not if we take one square root to be positive and the other square root to be negative. It is not understood why we are forced in this direction. We also know that, when $b = c = e = f = 0$, we should have $v_D\left(S_3\right) \neq 0$. If we take both square roots to be positive or both square roots to be negative, we have:

$$v_D\left(S_3\right) = \frac{\sinh\left(d\right)}{2} \qquad (13.47)$$

However, if we take one square root to be positive and the other square root to be negative, we have:

$$v_D\left(S_3\right) = -\frac{\sinh\left(d\right)}{6} \qquad (13.48)$$

We seem to get different rotations if we choose the signs of the square roots inconsistently. Consistency is next to godliness, and so choosing the square root signs consistently leads to the 2-dimensional S_3 rotation matrix:

$$S_3^{Rot}\Big|_{b=c=e=f=0} = \frac{1}{2} \begin{bmatrix} \cosh d & 0 & 0 & \sinh d & 0 & 0 \\ 0 & \cosh d & 0 & 0 & 0 & \sinh d \\ 0 & 0 & \cosh d & 0 & \sinh d & 0 \\ \sinh d & 0 & 0 & \cosh d & 0 & 0 \\ 0 & 0 & \sinh d & 0 & \cosh d & 0 \\ 0 & \sinh d & 0 & 0 & 0 & \cosh d \end{bmatrix}$$

(13.49)

This is not a double cover, but there is a strange $\frac{1}{2}$ involved. Of course, in some S_3 algebras, the hyperbolic trigonometric functions will be replaced by Euclidean trigonometric functions. In such spaces, because of the $\frac{1}{2}$, we will never be able to rotate around to the identity matrix, I, but will get to $\frac{1}{2}I$, which is where we started from.

Double cover in general:

We see that the question of double covers of the 2-dimensional spaces in our space-time is far from simple. The quaternions do simply and properly double cover the three 2-dimensional rotations in $SO(3)$. The 8-dimensional $C_2 \times C_2 \times C_2$ algebras do properly double cover the 2-dimensional rotations in $SO(4)$, but they do it only because one element of the algebra, the $e \equiv \overrightarrow{e_{123}}$ element, is 'special'. The special orthogonal group $SO(5)$ has ten 2-dimensional rotations within it (ten pairings of axes); it is not known if this is properly covered by the higher dimensional $C_2 \times C_2 \times ...$ algebras, and indeed, almost nothing of these things is known in dimensions higher than eight.

We see that the 6-dimensional S_3 spinors do not double cover any 2-dimensional rotations; after all, there is no $SO(3.5)$ special orthogonal group.

We can be sure that division algebras of odd dimensionality do not double cover 2-dimensional rotations because finite groups of odd order do not have order 2 sub-groups and so odd dimensional division algebras do not have 2-dimensional sub-algebras.

Chapter 14

Using Different Basis Elements as Generators

We have generated the Clifford algebras above from a small number of generators which we have taken to be the basis vectors $\vec{e}_i$. We proceeded thus:

1) Assume each basis vector is the square root of plus or minus unity.
2) Assume that the product of any two basis vectors is anti-commutative.

We have taken the view that a vector is not different from a bi-vector or any other multi-vector, and so we should be able to use basis elements other than the basis vectors to generate the algebra. There is a restriction upon our choice in that the basis elements we choose as generators must not form a sub-algebra for otherwise they would generate only the sub-algebra, but this is true of the basis vectors also. Let us consider $Cl_{3,0}$; we have:

The sub-algebras are:

8-dim Clifford algebra	$\sqrt{-1}$	$\sqrt{+1}$	$\mathbb{H}$	A_1	A_2	A_3
$Cl_{3,0}$ $\begin{pmatrix} 3\sqrt{+1} \\ 4\sqrt{-1} \end{pmatrix}$	4	3	$\left\{ \begin{array}{c} 1,\ \vec{e}_{12} \\ \vec{e}_{13},\ \vec{e}_{23} \end{array} \right\}$	0	$\left\{ \begin{array}{c} 1,\ \vec{e}_{1} \\ \vec{e}_{23},\ \vec{e}_{123} \end{array} \right\}$ $\left\{ \begin{array}{c} 1,\ \vec{e}_{2} \\ \vec{e}_{13},\ \vec{e}_{123} \end{array} \right\}$ $\left\{ \begin{array}{c} 1,\ \vec{e}_{3} \\ \vec{e}_{12},\ \vec{e}_{123} \end{array} \right\}$	$\left\{ \begin{array}{c} 1,\ \vec{e}_{1} \\ \vec{e}_{2},\ \vec{e}_{12} \end{array} \right\}$ $\left\{ \begin{array}{c} 1,\ \vec{e}_{1} \\ \vec{e}_{3},\ \vec{e}_{13} \end{array} \right\}$ $\left\{ \begin{array}{c} 1,\ \vec{e}_{2} \\ \vec{e}_{3},\ \vec{e}_{23} \end{array} \right\}$

The $Cl_{3,0}$ algebra is:

$$Cl_{3,0} \simeq d^2 = x^2 + y^2 + z^2$$
$$1, \quad \vec{e_1} = \sqrt{+1}, \quad \vec{e_2} = \sqrt{+1}, \quad \vec{e_3} = \sqrt{+1}, \tag{14.1}$$
$$\vec{e_{12}} = \sqrt{-1}, \quad \vec{e_{13}} = \sqrt{-1}, \quad \vec{e_{23}} = \sqrt{-1}, \quad \vec{e_{123}} = \sqrt{-1}$$

Suppose we choose the generators to be $\left\{ \vec{e_{12}}, \; \vec{e_{13}}, \; \vec{e_1} \right\}$; these do not all occur in a sub-algebra, and so they will generate an entire algebra. We assume these three generators are all the square roots of plus unity. This is not an exact copy of the $Cl_{3,0}$ algebra.

$$\vec{e_{12}}\,\vec{e_{12}} = +1, \quad \vec{e_{13}}\,\vec{e_{13}} = +1, \quad \vec{e_1}\,\vec{e_1} = +1 \tag{14.2}$$

We assume anti-commutation of these generators:

$$\vec{e_{12}}\,\vec{e_{13}} = -\vec{e_{13}}\,\vec{e_{12}}$$
$$\vec{e_{12}}\,\vec{e_1} = -\vec{e_1}\,\vec{e_{12}} \tag{14.3}$$
$$\vec{e_{13}}\,\vec{e_1} = -\vec{e_1}\,\vec{e_{13}}$$

From (14.2) and (14.3), after some calculation, we get:

$$\vec{e_2} = \sqrt{-1}, \quad \vec{e_3} = \sqrt{-1}, \quad \vec{e_{23}} = \sqrt{-1}, \quad \vec{e_{123}} = \sqrt{-1} \tag{14.4}$$

We have exactly the same relative numbers of square roots of plus unity and of square roots of minus unity as we have in the standard form of $Cl_{3,0}$. We note that, as with the standard form of $Cl_{3,0}$, the element $\vec{e_{123}}$ commutes with all other elements.

The above is not particularly interesting, but suppose we choose the a wholly commutative element, $\vec{e_{123}}$, to be a generator; it is only fair. If we make no distinction between vectors and multi-vectors, we can use a commutative basis element as a generator. This could not happen in the 4-dimensional Clifford algebras because the only commutative basis element was the real scalar which is a 1-dimensional sub-algebra and so cannot generate the entire algebra.

Choosing:

$$\overrightarrow{e_{12}}\,\overrightarrow{e_{12}} = +1, \quad \overrightarrow{e_1}\,\overrightarrow{e_1} = +1, \quad \overrightarrow{e_{123}}\,\overrightarrow{e_{123}} = +1 \tag{14.5}$$

We assume commutation and anti-commutation relations of these generators:

$$\overrightarrow{e_{12}}\,\overrightarrow{e_1} = -\overrightarrow{e_1}\,\overrightarrow{e_{12}}$$
$$\overrightarrow{e_{12}}\,\overrightarrow{e_{123}} = \overrightarrow{e_{123}}\,\overrightarrow{e_{12}} \tag{14.6}$$
$$\overrightarrow{e_1}\,\overrightarrow{e_{123}} = \overrightarrow{e_{123}}\,\overrightarrow{e_1}$$

From (14.5) and (14.6), it follows that:

$$\overrightarrow{e_2} = \sqrt{-1}$$
$$\overrightarrow{e_3} = \sqrt{+1}$$
$$\overrightarrow{e_{23}} = \sqrt{+1} \tag{14.7}$$
$$\overrightarrow{e_{13}} = \sqrt{-1}$$

The $Cl_{3,0}$ algebra is of the form $\left(3\sqrt{+1}, 4\sqrt{-1}\right)$. Immediately above, we have $\left(5\sqrt{+1}, 2\sqrt{-1}\right)$ which is the form of $Cl_{2,1}$. Because we chose a commutative basis element to be one of the generators of the algebra, we arrived at a different algebra from when the three generators were all anti-commutative.

A shortfall of the Clifford algebras:

The conventional construction of the Clifford algebras assumes that all the generating elements of the algebra are anti-commutative. We could extend the construction to allow some of the generating elements to be commutative. This would lead to other algebras. In two, four, and eight dimensions, the other algebras are already known Clifford algebras that could have been calculated by the conventional construction. In sixteen dimensions, as we will see shortly, the conventional construction 'misses' some algebras. This means that the neat one-to-one correspondence between $C_2 \times C_2 \times...$ division algebras and the Clifford algebras breaks down for algebras of dimension greater than eight. This is no great mathematical collapse; it is a weakness of the conventional way of constructing Clifford algebras.

The $C_2 \times C_2 \times C_2 \times C_2$ Division Algebras:

It is a direct consequence of the standard Clifford algebra theorem, (5.30):

$$Cl_{p,q} \simeq Cl_{q+1,p-1}$$
$$Cl_{p,q} \simeq Cl_{p-4,q+4}$$

(15.1)

that there are only two algebraically distinct 16-dimensional Clifford algebras. These are:

$$Cl_{4,0} \sim Cl_{0,4} \sim Cl_{1,3} \sim \left\{1, 5\sqrt{+1}, 10\sqrt{-1}\right\}$$
$$Cl_{3,1} \sim Cl_{2,2} \sim \left\{1, 9\sqrt{+1}, 6\sqrt{-1}\right\}$$

(15.2)

There are five distinct 16-dimensional non-commutative $C_2 \times C_2 \times ...$ division algebras with the following ratios of imaginary square roots of plus unity and square roots of minus unity:

$\sqrt{+1}$	$\sqrt{-1}$
3	12
5	10
7	8
9	6
11	4

(15.3)

We have three more types of division algebras than we have types of Clifford algebras.

The Clifford algebra $Cl_{4,0}$:

The Clifford algebra $Cl_{4,0}$ has the basis elements:

$$Cl_{4,0} \simeq$$

$$1, \ \vec{e_1} = \sqrt{+1}, \ \vec{e_2} = \sqrt{+1}, \ \vec{e_3} = \sqrt{+1}, \ \vec{e_4} = \sqrt{+1},$$

$$\vec{e_{12}} = \sqrt{-1}, \ \vec{e_{13}} = \sqrt{-1}, \ \vec{e_{14}} = \sqrt{-1}, \ \vec{e_{23}} = \sqrt{-1}, \ \vec{e_{24}} = \sqrt{-1}, \ \vec{e_{34}} = \sqrt{-1},$$

$$\vec{e_{123}} = \sqrt{-1}, \ \vec{e_{124}} = \sqrt{-1}, \ \vec{e_{134}} = \sqrt{-1}, \ \vec{e_{234}} = \sqrt{-1}, \ \vec{e_{1234}} = \sqrt{+1}$$

$$(15.4)$$

The anti-commutation relations of this algebra are:

$$\vec{e_1}\vec{e_2} = -\vec{e_2}\vec{e_1} \qquad \vec{e_1}\vec{e_3} = -\vec{e_3}\vec{e_1}$$

$$\vec{e_1}\vec{e_4} = -\vec{e_4}\vec{e_1} \qquad \vec{e_2}\vec{e_3} = -\vec{e_3}\vec{e_2} \qquad (15.5)$$

$$\vec{e_2}\vec{e_4} = -\vec{e_4}\vec{e_2} \qquad \vec{e_3}\vec{e_4} = -\vec{e_4}\vec{e_3}$$

The centre of this algebra, $Cl_{4,0}$, is only one element, $\{1\}$. There is only one element that anti-commutes with all other elements; that is the $\overline{e_{1234}}$ element. The same is true for the $Cl_{3,1}$ algebra. If we are to use the basis vectors to generate the algebra and these basis vectors are all anti-commutative, then it will always be the case that 'shifting a basis vector' by one position, left or right, changes the sign of the product; for example:

$$\vec{e_2}\overline{e_{1234}} = -\overline{e_{12234}}$$

$$\overline{e_{1234}}\vec{e_2} = -\vec{e_2}\overline{e_{1234}} = +\overline{e_{12234}} \qquad (15.6)$$

$$\vec{e_2}\overline{e_{1234}} = -\overline{e_{1234}}\vec{e_2}$$

This is true in all such algebras which differ only in the relative numbers of square roots of minus unity and square roots of plus unity. Thus all Clifford algebras of a given dimension will have the same centre and the same wholly anti-commutative variables.

With a little thought, we might realise that the $\overline{e_{12345}}$ element of the 32 dimensional Clifford algebras will be fully commutative and the

$\overline{e_{123456}}$ element of the 64 dimensional Clifford algebra will be fully anti-commutative etc..

Since in 16 dimensions we have no fully commutative variable other than the identity, we cannot form a different algebra by using a commutative basis element as a generator of the algebra. This, in 16 dimensions, is where conventional Clifford algebra grinds to a halt; division algebras do not halt here.

The $5\sqrt{+1}$, $10\sqrt{-1}$ division algebra and the $9\sqrt{+1}$, $6\sqrt{-1}$ division algebras all have a centre of only the identity $\{a\}$. This corresponds to the two types of Clifford algebra. The $5\sqrt{+1}$, $10\sqrt{-1}$ algebra occurs in 43,008 different bases. The $9\sqrt{+1}$, $6\sqrt{-1}$ algebra occurs in 71,680 different bases.

The remaining division algebras:

The non-commutative division algebras $7\sqrt{+1}$, $8\sqrt{-1}_{non-com}$, $3\sqrt{+1}$, $12\sqrt{-1}$ and $11\sqrt{+1}$, $4\sqrt{-1}$ each have a centre of four elements; that is the identity and three imaginary elements are commutative. You see, we have more non-commutative division algebras than we have Clifford algebras. We see that the number of elements in the centre is the defining difference.

By the way, there are two fully commutative 16-dimensional division algebras. These are $7\sqrt{+1}$, $8\sqrt{-1}_{com}$ and $15\sqrt{+1}$.

Chapter 16

Questionable Interpretation of Clifford Algebras

Above, we have seen that the Clifford algebras are algebraically the same as the $C_2 \times C_2 \times ...$ division algebras, but there are subtle interpretational difficulties within the conventional view of Clifford algebras that make them different from division algebras. The n-dimensional $C_2 \times C_2 \times ...$ division algebras are comprised of one real variable and $(n-1)$ imaginary variables. Conventionally, we construct a Clifford algebra from a number of real basis vectors, $\vec{e_1}, \vec{e_2}, ...$, but we then make these basis vectors the square roots of plus unity or the square roots of minus unity, and so we have real elements that are the square roots of unity or the square roots of minus unity. These basis vectors are something different from unity, and, other than unity itself, there is no such thing as a real square root of plus unity or minus unity; we expect these basis vectors to be real elements and imaginary elements at the same time. The confusion arises from confusing $\mathbb{R}^n$ space with division algebra space. An example is $Cl_{2,0}$ which is conventionally taken to be based on one real scalar, two real basis vectors, and one imaginary bi-vector whereas the division algebra associated with this Clifford algebra has three imaginary variables and one real variable. It seems to your author, that the conventional view of Clifford algebra is flawed in this sense.

The norms of division algebras and Clifford algebras:
Conventionally, Clifford algebras are taken to have norms that are quadratic forms; for example:

$$Cl_{2,0} \sim x^2 + y^2 \qquad Cl_{1,1} \sim x^2 - y^2$$
$$Cl_{3,0} \sim x^2 + y^2 + z^2 \qquad Cl_{1,2} \sim x^2 - y^2 - z^2 \qquad (16.1)$$
$$Cl_{4,0} \sim t^2 + x^2 + y^2 + z^2 \qquad Cl_{1,3} \sim t^2 - x^2 - y^2 - z^2$$

We have seen above, (12.6), that the Clifford algebra $Cl_{3,0}$ is the same as the 8-dimensional $\left\{1,\ 3\sqrt{+1},\ 4\sqrt{-1}\right\}$ division algebra. The norm of this 8-dimensional algebra is not a quadratic form and it has eight variables. The norm conventionally associated with a Clifford algebra contains only the basis vector variables; this is nothing like the norm of the associated division algebra that contains all the variables in the algebra and is not necessarily a quadratic form. In this sense, the Clifford algebras differ from the division algebras. Your author takes the view that the Clifford algebras are flawed also in this sense.

Chapter 17

Sub-algebras

Even sub-algebras:
Within a Clifford algebra, there is a sub-algebra known as the even sub-algebra. The even sub-algebra is comprised of all the basis elements of the Clifford algebra that are even multiples of the basis vectors plus the scalar element[82]. For example, within $Cl_{2,0}$ there are four basis elements $\{1, \vec{e_1}, \vec{e_2}, \vec{e_{12}}\}$; of these basis elements $\{1, \vec{e_{12}}\}$ form the even sub-algebra. In the case of $Cl_{2,0}$, this even sub-algebra is algebraically isomorphic to the complex numbers $\mathbb{C}$, and we have $\{1 \sim 1,\ \vec{e_{12}} \sim i\}$. The even sun-algebra is written with a superscript plus sign:

$$Cl_{2,0}^{+} = \{1,\ \vec{e_{12}}\} = \{1, \sqrt{-1}\} \qquad (17.1)$$

Another example is the even sub-algebra of $Cl_{3,0}$ which is isomorphic to the quaternions, $\mathbb{H}$:

$$Cl_{3,0}^{+} = \{1,\ \vec{e_{12}},\ \vec{e_{13}},\ \vec{e_{23}}\} = \{1, \sqrt{-1}, \sqrt{-1}, \sqrt{-1}\} \qquad (17.2)$$

List of even sub-algebras:
We list the even sub-algebras of the lower dimensional Clifford algebras. We include algebras like the A_3 algebras which are not conventionally considered to be even sub-algebras because they become division algebras in only their polar forms.

---

[82] The scalar element is a real number that is viewed as being zero multiples of the basis vectors and zero is viewed as being an even number.

			Isomorphic to:
$Cl_{1,0}^{+}$	1	1	$\mathbb{R}$
$Cl_{0,1}^{+}$	1	1	$\mathbb{R}$
$Cl_{2,0}^{+}$	$\{1, \vec{e}_{12}\}$	$\{1, \sqrt{-1}\}$	$\mathbb{C}$
$Cl_{1,1}^{+}$	$\{1, \vec{e}_{12}\}$	$\{1, \sqrt{+1}\}$	$\mathbb{S}$ [83]
$Cl_{0,2}^{+}$	$\{1, \vec{e}_{12}\}$	$\{1, \sqrt{-1}\}$	$\mathbb{C}$

The even sub-algebra $Cl_{1,1}^{+}$ is not quite isomorphic to the hyperbolic complex numbers unless we take the exponential, but we include it for completeness.

			Isomorphic to:
$Cl_{3,0}^{+}$	$\{1, \vec{e}_{12}, \vec{e}_{13}, \vec{e}_{23}\}$	$\{1, \sqrt{-1}, \sqrt{-1}, \sqrt{-1}\}$	$\mathbb{H}$
$Cl_{2,1}^{+}$	$\{1, \vec{e}_{12}, \vec{e}_{13}, \vec{e}_{23}\}$	$\{1, \sqrt{-1}, \sqrt{+1}, \sqrt{+1}\}$	A_3
$Cl_{1,2}^{+}$	$\{1, \vec{e}_{12}, \vec{e}_{13}, \vec{e}_{23}\}$	$\{1, \sqrt{+1}, \sqrt{+1}, \sqrt{-1}\}$	A_3
$Cl_{0,3}^{+}$	$\{1, \vec{e}_{12}, \vec{e}_{13}, \vec{e}_{23}\}$	$\{1, \sqrt{-1}, \sqrt{-1}, \sqrt{-1}\}$	$\mathbb{H}$

			Isomorphic to:
$Cl_{4,0}^{+}$	$\left\{ \begin{array}{c} 1, \vec{e}_{12}, \vec{e}_{13}, \vec{e}_{14} \\ \vec{e}_{23}, \vec{e}_{24}, \vec{e}_{34}, \vec{e}_{1234} \end{array} \right\}$	$\left\{ \begin{array}{c} 1, \sqrt{-1}, \sqrt{-1}, \sqrt{-1} \\ \sqrt{-1}, \sqrt{-1}, \sqrt{-1}, \sqrt{+1} \end{array} \right\}$	$\{1, \sqrt{+1}, 6\sqrt{-1}\}$

---

[83] $\mathbb{S}$ is the hyperbolic complex numbers

$Cl_{0,4}^+$	$\left\{ \begin{array}{l} 1,\ \vec{e}_{12},\ \vec{e}_{13},\ \vec{e}_{14} \\ \vec{e}_{23},\ \vec{e}_{24},\ \vec{e}_{34},\ \vec{e}_{1234} \end{array} \right\}$	$\left\{ \begin{array}{l} 1,\ \sqrt{-1},\ \sqrt{-1},\ \sqrt{-1} \\ \sqrt{-1},\ \sqrt{-1},\ \sqrt{-1},\ \sqrt{+1} \end{array} \right\}$	$\left\{ 1,\ \sqrt{+1},\ 6\sqrt{-1} \right\}$
$Cl_{1,3}^+$	$\left\{ \begin{array}{l} 1,\ \vec{e}_{12},\ \vec{e}_{13},\ \vec{e}_{14} \\ \vec{e}_{23},\ \vec{e}_{24},\ \vec{e}_{34},\ \vec{e}_{1234} \end{array} \right\}$	$\left\{ \begin{array}{l} 1,\ \sqrt{+1},\ \sqrt{+1},\ \sqrt{+1} \\ \sqrt{-1},\ \sqrt{-1},\ \sqrt{-1},\ \sqrt{-1} \end{array} \right\}$	$\left\{ 1,\ 3\sqrt{+1},\ 4\sqrt{-1} \right\}$
$Cl_{3,1}^+$	$\left\{ \begin{array}{l} 1,\ \vec{e}_{12},\ \vec{e}_{13},\ \vec{e}_{14} \\ \vec{e}_{23},\ \vec{e}_{24},\ \vec{e}_{34},\ \vec{e}_{1234} \end{array} \right\}$	$\left\{ \begin{array}{l} 1,\ \sqrt{-1},\ \sqrt{-1},\ \sqrt{+1} \\ \sqrt{-1},\ \sqrt{+1},\ \sqrt{+1},\ \sqrt{-1} \end{array} \right\}$	$\left\{ 1,\ 3\sqrt{+1},\ 4\sqrt{-1} \right\}$
$Cl_{2,2}^+$	$\left\{ \begin{array}{l} 1,\ \vec{e}_{12},\ \vec{e}_{13},\ \vec{e}_{14} \\ \vec{e}_{23},\ \vec{e}_{24},\ \vec{e}_{34},\ \vec{e}_{1234} \end{array} \right\}$	$\left\{ \begin{array}{l} 1,\ \sqrt{-1},\ \sqrt{+1},\ \sqrt{+1} \\ \sqrt{+1},\ \sqrt{+1},\ \sqrt{-1},\ \sqrt{+1} \end{array} \right\}$	$\left\{ 1,\ 5\sqrt{+1},\ 2\sqrt{-1} \right\}$

We see that the even sub-algebras of the Clifford algebras are isomorphic to division algebras of the appropriate dimension.

Other sub-algebras:

Within conventional Clifford algebra, because we distinguish between basis vectors and basis bi-vectors etc., the even sub-algebra is distinguished from other sub-algebras. Within the division algebras, because we do not distinguish between basis vectors and basis bi-vectors etc., there is nothing to distinguish between one sub-algebra and any other sub-algebra of the same dimension. We see that there is no distinction between even basis elements and odd basis elements within a Clifford algebra.

Traditionally, Clifford algebraists have been interested in only the sub-algebras of Clifford algebras that are isomorphic to the complex numbers, $\mathbb{C}$, or the quaternions, $\mathbb{H}$. If we were unable to take the exponential of an algebraic matrix form to rid us of singular matrices, the only division algebras would be $\{\mathbb{R}, \mathbb{C}, \mathbb{H}\}$[84]. All the other division algebras are, like the hyperbolic complex numbers, 'not everywhere' but exist only between some asymptotes such as the limiting velocity of the universe. Of course, Clifford algebraists traditionally do not take the exponential, and so the algebras other than $\{\mathbb{R}, \mathbb{C}, \mathbb{H}\}$ 'do not exist' in the Clifford algebraist's view.

If we are to identify Clifford algebras with the division algebras, we must not distinguish between even sub-algebras and other sub-algebras. Spinors are traditionally associated with the even sub-algebras of Clifford algebras. It seems that we must associate spinors with all sub-algebras; but a Clifford sub-algebra is a Clifford algebra in its own right; we will associate spinors with all Clifford algebras.

Spinors again:
So, instead of spinors existing in only some Clifford algebras and then in only the even sub-algebras, spinors now exist in all Clifford algebras. We identify Clifford algebras with the $C_2 \times C_2 \times ...$ division algebras, and so spinors exist in all the $C_2 \times C_2 \times ...$ division algebras. They are the rotation matrices of these algebras. Well, it is unfair to favour only the $C_2 \times C_2 \times ...$ division algebras, and so we generalise. Spinors exist in all division algebras, but they come in different varieties; some have double cover and some do not. It seems that only the spinors in the $C_2 \times C_2 \times ...$ division algebras are involved in physics. Most spinors are within commutative division algebras and play no part in physics.

---

[84] The octonians are non-associative, and so we do not count them as a division algebra.

Chapter 18

Differentiation in Clifford Algebras

Geometric algebra and geometric calculus:
Clifford algebra is today often called geometric algebra[85]. Although Clifford algebra traditionally does not include differentiation, such calculus has been developed and is called geometric calculus. The conventional geometric derivative is[86]:

$$\nabla = e^i \partial_i \qquad (18.1)$$

There is summation over the values of i. This derivative is similar to the gradient but acts upon multi-vector valued functions, $F(a)$ rather than acting on only vectors. However, there are problems with the geometric derivative of a product due to the non-commutative nature of Clifford algebras (geometric algebras). None-the-less, it is possible to define both an interior derivative and an exterior derivative of $F(a)$ provided $F(a)$ is comprised of only elements of the Clifford algebra of the same type (all vectors or all bi-vectors or all tri-vectors or …). The interior derivative and exterior derivative are respectfully:

$$\nabla \cdot F = e^i \cdot \partial_i F$$
$$\nabla \wedge F = e^i \wedge \partial_i F \qquad (18.2)$$

If $F(a)$ is vector valued (no bi-vectors etc.), we have[87]:

$$\nabla F = \nabla \cdot F + \nabla \wedge F \qquad (18.3)$$

---

[85] Clifford algebraists do love their geometric interpretation.
[86] Geometric calculus: Wikipedia
[87] This is how differentiation is defined by David Hestenes in 'Vectors, Spinors, and Complex Numbers in Classical and Quantum Physics: American Journal of Physics Vol 39/9 1013-1027 September 1971

Neither the interior derivative nor the exterior derivative is invertible. We follow Lounesto[88]. The differential of a vector:

$$\vec{E} = E_x \vec{e_1} + E_y \vec{e_2} + E_z \vec{e_3} \tag{18.4}$$

in $Cl_{3,0}$ is:

$$\nabla \vec{E} \equiv \overline{\overline{\nabla \vec{E}}} = \nabla \cdot \vec{E} + \nabla \wedge \vec{E} \tag{18.5}$$

In $Cl_{3,0}$, we have:

$$\left(\frac{\partial}{\partial x} \vec{e_1} + \frac{\partial}{\partial y} \vec{e_2} + \frac{\partial}{\partial z} \vec{e_3} \right) \left(E_x \vec{e_1} + E_y \vec{e_2} + E_z \vec{e_3} \right)$$

$$= \frac{\partial E_x}{\partial x} \vec{e_{11}} + \frac{\partial E_y}{\partial x} \vec{e_{12}} + \frac{\partial E_z}{\partial x} \vec{e_{13}}$$

$$+ \frac{\partial E_x}{\partial y} \vec{e_{21}} + \frac{\partial E_y}{\partial y} \vec{e_{22}} + \frac{\partial E_z}{\partial y} \vec{e_{23}} \tag{18.6}$$

$$+ \frac{\partial E_x}{\partial z} \vec{e_{31}} + \frac{\partial E_y}{\partial z} \vec{e_{32}} + \frac{\partial E_z}{\partial z} \vec{e_{33}}$$

Rearranging:

$$\nabla \vec{E} = \left(\frac{\partial E_x}{\partial x} + \frac{\partial E_y}{\partial y} + \frac{\partial E_z}{\partial z} \right)$$

$$+ \left(\frac{\partial E_y}{\partial x} - \frac{\partial E_x}{\partial y} \right) \vec{e_{12}} + \left(\frac{\partial E_z}{\partial x} - \frac{\partial E_x}{\partial z} \right) \vec{e_{13}} \tag{18.7}$$

$$+ \left(\frac{\partial E_z}{\partial y} - \frac{\partial E_y}{\partial z} \right) \vec{e_{23}}$$

This is the divergence and it is a bi-vector. To get from the bi-vector sub-space back into the vector sub-space, we need to use the Hodge dual[89]:

---

[88] Lounesto: page 107
[89] Lounesto: page 39 footnotes

$$\vec{a} \times \vec{b} = -\left(\vec{a} \wedge \vec{b}\right)\overrightarrow{e_{123}} \qquad (18.8)$$

We multiply the bi-vectors by $\overrightarrow{e_{123}}$ on the right and reverse the sign to get:

$$
\begin{aligned}
\nabla\vec{E} =& \left(\frac{\partial E_x}{\partial x} + \frac{\partial E_y}{\partial y} + \frac{\partial E_z}{\partial z}\right) - \left(\frac{\partial E_y}{\partial x} - \frac{\partial E_x}{\partial y}\right)\overrightarrow{e_{12}}\,\overrightarrow{e_{123}} \\
& -\left(\frac{\partial E_z}{\partial x} - \frac{\partial E_x}{\partial z}\right)\overrightarrow{e_{13}}\,\overrightarrow{e_{123}} - \left(\frac{\partial E_z}{\partial y} - \frac{\partial E_y}{\partial z}\right)\overrightarrow{e_{23}}\,\overrightarrow{e_{123}} \\
=& \left(\frac{\partial E_x}{\partial x} + \frac{\partial E_y}{\partial y} + \frac{\partial E_z}{\partial z}\right) + \left(\frac{\partial E_y}{\partial x} - \frac{\partial E_x}{\partial y}\right)\overrightarrow{e_3} \\
& -\left(\frac{\partial E_z}{\partial x} - \frac{\partial E_x}{\partial z}\right)\overrightarrow{e_2} + \left(\frac{\partial E_z}{\partial y} - \frac{\partial E_y}{\partial z}\right)\overrightarrow{e_3}
\end{aligned}
\qquad (18.9)
$$

Thus we have differentiated a vector in $Cl_{3,0}$ and got both the divergence and the curl of that vector in one operation. Well, we have, but the introduction of the tri-vector to return us to the vector sub-space seems a little contrived.

Differentiation in division algebras:
Of course, within the division algebra spaces, we do not distinguish between vectors and multi-vectors and differentiation is properly defined as matrix differentiation in both commutative and non-commutative forms[90]. Within the division algebras that correspond to Clifford algebras, we simply differentiate non-commutatively, and this effectively gives the same result as we gained above, (18.9), in a straight-forward way. Of course, within the division algebras we cannot work in $\mathbb{R}^3$ but must work in the entire algebra; we cannot reproduce exactly (18.9) but instead get the above as part of the 4-dimensional derivative – we differentiate a 4-dimensional vector. It is remarkable that the clearly defined non-commutative differentiation

---

[90] See: Dennis Morris The Physics of Empty Space. ISBN: 978-1-5077-0700-5

within division algebras is the same thing as differentiation in the geometric calculus.

Thus, it seems to your author, the easiest way to differentiate within a Clifford algebra is to convert the Clifford algebra to a division algebra and differentiate within the division algebra. Of course, it seems to your author, that the easiest way to do Clifford algebra in general is to do division algebra instead.

Matrix differentiation:
We will differentiate a complex function with respect to an imaginary variable to show the technique[91]:

$$
\frac{\partial \begin{bmatrix} f(x,y) & g(x,y) \\ -g(x,y) & f(x,y) \end{bmatrix}}{\partial \begin{bmatrix} 0 & x \\ -x & 0 \end{bmatrix}} = \frac{\partial \begin{bmatrix} f(x,y) & 0 \\ 0 & f(x,y) \end{bmatrix}}{\begin{bmatrix} 0 & 1 \\ -1 & 0 \end{bmatrix} \partial \begin{bmatrix} x & 0 \\ 0 & x \end{bmatrix}}
$$

$$
+ \frac{\begin{bmatrix} 0 & 1 \\ -1 & 0 \end{bmatrix} \partial \begin{bmatrix} g(x,y) & 0 \\ 0 & g(x,y) \end{bmatrix}}{\begin{bmatrix} 0 & 1 \\ -1 & 0 \end{bmatrix} \partial \begin{bmatrix} x & 0 \\ 0 & x \end{bmatrix}}
$$

$$
= \begin{bmatrix} 0 & -1 \\ 1 & 0 \end{bmatrix} \begin{bmatrix} \dfrac{\partial f}{\partial x} & 0 \\ 0 & \dfrac{\partial f}{\partial x} \end{bmatrix} + \begin{bmatrix} \dfrac{\partial g}{\partial x} & 0 \\ 0 & \dfrac{\partial g}{\partial x} \end{bmatrix} = \begin{bmatrix} \dfrac{\partial g}{\partial x} & -\dfrac{\partial f}{\partial x} \\ \dfrac{\partial f}{\partial x} & \dfrac{\partial g}{\partial x} \end{bmatrix}
$$

$$(18.10)$$

All we do is extract the imaginary unit variables and then differentiate as we do with real numbers before putting the imaginary unit variables back into the answer.

Above, (18.10), we extracted the imaginary unit variables to the left. Since the complex numbers are commutative, it matters not to which

---

[91] See: Dennis Morris The Physics of Empty Space. ISBN: 978-1-507707-00-5

side we extract the imaginary unit variables in this division algebra. However, in a non-commutative division algebra such as the quaternions, we get two different differentials depending upon whether we extract the imaginary unit variables to the left or to the right. We call these the left differential, d_L, and the right differential, d_R, respectively. When we do this, we differentiate with respect to all the individual variables (real and imaginary) and add the results to get the left differential and the right differential. We then form the two differential fields:

$$E = \frac{1}{2}(d_L + d_R)$$
$$B = \frac{1}{2}(d_L - d_R)$$

(18.11)

These fields are so named because when we differentiate a quaternion potential[92], we get the electric and magnetic fields. Differentiation of an anti-quaternion gives the anti-electric and the anti-magnetic fields. Superimposition of these four fields gives the classical electromagnetic tensor in quaternion format (distribution of minus signs) – no anti-matter. A second differentiation followed by superimposition gives the Maxwell equations – no need to assume a lagrangian.

The reader might want to check that non-commutative differentiation of a quaternion potential gives the same answer as the conventional Clifford differentiation. There are no problems with matrix differentiation; everything works, and so we prefer it to conventional Clifford differentiation.

---

[92] We need to use the conjugate quaternion potential to correspond with the arbitrary definitions of the electric and magnetic fields.

Chapter 19

Rotation and Spinors in Clifford Algebras

This chapter will be easier to understand if the reader keeps in mind the nature of a division algebra rotation matrix. To that end, we present the quaternion rotation matrix; this is equivalent to the $SU(2)$ spinor representation of the Lie group $SO(3)$:

$$\mathbb{H}_{Rot} = \begin{bmatrix} \cos(\lambda) & \dfrac{b}{\lambda}\sin(\lambda) & \dfrac{c}{\lambda}\sin(\lambda) & \dfrac{d}{\lambda}\sin(\lambda) \\[3mm] -\dfrac{b}{\lambda}\sin(\lambda) & \cos(\lambda) & -\dfrac{d}{\lambda}\sin(\lambda) & \dfrac{c}{\lambda}\sin(\lambda) \\[3mm] -\dfrac{c}{\lambda}\sin(\lambda) & \dfrac{d}{\lambda}\sin(\lambda) & \cos(\lambda) & -\dfrac{b}{\lambda}\sin(\lambda) \\[3mm] -\dfrac{d}{\lambda}\sin(\lambda) & -\dfrac{c}{\lambda}\sin(\lambda) & \dfrac{b}{\lambda}\sin(\lambda) & \cos(\lambda) \end{bmatrix} \quad (19.1)$$

$$\lambda = \sqrt{b^2 + c^2 + d^2}$$

Notice that the quaternion trigonometric functions accept three arguments (angles $\{b,c,d\}$). This multi-argument (multi-angle) nature of the trigonometric functions is one of the differences between spinor rotation and rotation in our 4-dimensional space-time. Rotation in our 4-dimensional space-time is a sequence of 2-dimensional rotations. Of course, the 2-dimensional trigonometric functions each take only one argument (angle). Also recall that quaternion rotation is not rotation about an axis; this is another difference between spinor rotation and rotation in our 4-dimensional space-time.

Here is the rub; there are different types of geometric space; for example, 4-dimensional quaternion space is different from the 4-dimensional space-time in which we sit. Since there are different types of geometric space, there are different types of rotation. Rotation in division algebra spaces is spinor rotation. Rotation in $\mathbb{R}^n : n > 2$ spaces is 'normal' rotation. Since we opine that division algebra spaces are quantum spaces, we associate spinor rotation with quantum physics. Since we opine that $\mathbb{R}^n : n > 2$ spaces are classical spaces, we associate 'normal' rotation with classical physics.

Two types of rotation imply two types of angular momentum. We associate rotation in classical spaces with orbital angular momentum, and we associate rotation within division algebra spaces with intrinsic spin.

On top of all this, to add to the confusion, there are classical Lie groups which are often taken to represent rotations in various types of 'classical' space but are actually rotations in various invented linear spaces that preserve the invented inner products placed upon those linear spaces by humans. A linear space is not generally a geometric space.

Perceived rotation and expressions for rotation:

There are different mathematical ways of expressing rotation even within a single type of space. We sometimes see rotation expressed with a single rotation matrix, and we sometimes see rotation expressed with a pair of rotation matrices, and we sometimes see rotation expressed seemingly without any rotation matrices. We will deal with this notational cacophony shortly.

Of course, we humans think of rotation as a continuous movement around a circle or along a geodesic (great circle) upon the surface of a sphere. Mathematically, there is no continuity between the starting point, A, of a rotation and the finishing point, B, of that rotation; we simply multiply by a rotation matrix or two and at no point in that multiplicative procedure can we say we have now done part of the multiplication and so we have rotated part of the way from the starting point to the end point. The act of multiplication is a single 'instantaneous' operation that changes point A into point B with no intermediate points being visited.

Rotation in $\mathbb{R}^n$ types of space:

Mathematicians are perhaps most familiar with rotation in spaces constructed of copies of the real numbers usually referred to as $\mathbb{R}^n$ space. Within these spaces, all rotations are 2-dimensional and are associated with one of two possible 2×2 rotation matrices:

$$\begin{bmatrix} \cos\theta & \sin\theta \\ -\sin\theta & \cos\theta \end{bmatrix} \quad \text{or} \quad \begin{bmatrix} \cosh\chi & \sinh\chi \\ \sinh\chi & \cosh\chi \end{bmatrix} \qquad (19.2)$$

I'll just say that again; in $\mathbb{R}^n$ space, all rotations are 2-dimensional. Perhaps the reader is now thinking she was foolish to begin reading this book; it seems that the author is insane; what kind of rotation is not 2-dimensional? We sit in a 4-dimensional space; 'obviously', rotation should be 4-dimensional similar to rotation in quaternion space – see (19.1) – that's the sane view. So why do we have only 2-dimensional rotations in our 4-dimensional space-time?

The leftmost of the above rotations, (19.2), is a Euclidean rotation and the rightmost is often called a boost because a rotation in space-time is a change of velocity. Such 2-dimensional rotations vary with a single parameter which we call the angle - θ & χ above, (19.2). In $\mathbb{R}^n$ space, there are as many such 2-dimensional rotations as there are pairs of axes; for example, there are three Euclidean rotations and three boosts in the 4-dimensional space-time in which we sit corresponding to three ways of pairing together spatial axes and three pairs of the time axis with a spatial axis respectively.

These $\mathbb{R}^n$ spaces can have 2-dimensional rotation(s) within them only if the n-dimensional distance function of the n-dimensional space accommodates (includes) the 2-dimensional distance functions associated with the appropriate 2-dimensional rotation matrices. The 2-dimensional distance function is, after all, what a 2-dimensional rotation leaves invariant. Those 2-dimensional distance functions are both quadratic forms:

$$d^2 = x^2 + y^2 \quad \text{or} \quad d^2 = t^2 - z^2 \qquad (19.3)$$

We see that both of these 2-dimensional quadratic forms are accommodated within the 4-dimensional distance function of space-time:

$$d^2 = t^2 - x^2 - y^2 - z^2 \qquad (19.4)$$

Thus it is that the $\mathbb{R}^n$ spaces that have 2-dimensional rotations must have quadratic forms for (at least part of) the distance function. Above, we saw that the $\mathbb{R}^n$ spaces derive from the division algebra spaces by superimposition and that the 2-dimensional rotations survive the superimposition operation. Even so, unless the distance function of the derived $\mathbb{R}^n$ space has within it the 2-dimensional distance functions, there cannot be 2-dimensional rotations that preserve that distance function within that space. An example is the 3-dimensional $\mathbb{R}^n$ space that cannot support 2-dimensional rotations.

Aside: The reader might have noticed that the two 2-dimensional rotation matrices are rotation matrices within division algebras; as such these 2-dimensional rotations are spinor rotations. In 2-dimensional space, we have only spinor rotations. There are no 2-dimensional spaces of the form $\mathbb{R}^2$ because the 2-dimensional algebras survive superimposition.

Eigenvectors and $\mathbb{R}^n$ rotations:

Consider the 2-dimensional rotation matrix in $\mathbb{R}^3$:

$$R = \begin{bmatrix} \dfrac{\sqrt{3}}{2} & -\dfrac{1}{2} & 0 \\[2mm] \dfrac{1}{2} & \dfrac{\sqrt{3}}{2} & 0 \\[2mm] 0 & 0 & 1 \end{bmatrix} \qquad (19.5)$$

This represents a rotation through $\dfrac{\pi}{6}$ radians about the z-axis. The eigenvalues of this matrix, (19.5), are:

145

$$\lambda_1 = 1, \quad \lambda_2 = \frac{\sqrt{3}}{2} + i\frac{1}{2} = e^{i\frac{\pi}{6}}, \quad \lambda_3 = \frac{\sqrt{3}}{2} - i\frac{1}{2} = e^{-i\frac{\pi}{6}} \quad (19.6)$$

These eigenvalues correspond respectively to the eigenvectors:

$$\begin{bmatrix} 0 \\ 0 \\ 1 \end{bmatrix}, \quad \begin{bmatrix} i \\ 1 \\ 0 \end{bmatrix}, \quad \begin{bmatrix} -i \\ 1 \\ 0 \end{bmatrix} \quad (19.7)$$

The eigenvalue $\lambda_1 = 1$ and the corresponding eigenvector show no change in the z-direction by this rotation. In other words, this is a rotation about the z-axis. In general, the eigenvalues of:

$$R_G = \begin{bmatrix} \cos\theta & \sin\theta & 0 \\ -\sin\theta & \cos\theta & 0 \\ 0 & 0 & 1 \end{bmatrix} \quad (19.8)$$

are:

$$\cos\theta + i\sin\theta, \quad \cos\theta - i\sin\theta, \quad 1 \quad (19.9)$$

The 1 corresponds to a direction that is unchanged by the rotation – the z-axis.

Rotation within a division algebra space:

The $\mathbb{R}^n$ spaces are not the only type of geometric spaces. There are also n-dimensional spaces which have one real axis and $(n-1)$ imaginary axes. These are the division algebra spaces; examples are the complex plane, $\mathbb{C}$, the hyperbolic complex plane (2-dimensional space-time), $\mathbb{S}$, or the quaternions, $\mathbb{H}$. In division algebra spaces, we associate rotation with a rotation matrix full of trigonometric functions[93] (no zeros or 1's allowed); for example, rotation in the complex plane, $\mathbb{C}$, of a point (a,b) is given by:

$$\begin{bmatrix} \cos\theta & \sin\theta \\ -\sin\theta & \cos\theta \end{bmatrix}\begin{bmatrix} a & b \\ -b & a \end{bmatrix} \tag{19.10}$$

Rotation in the hyperbolic complex plane, $\mathbb{S}$, of a point (t,z) is given by:

$$\begin{bmatrix} \cosh\chi & \sinh\chi \\ \sinh\chi & \cosh\chi \end{bmatrix}\begin{bmatrix} t & z \\ z & t \end{bmatrix} \tag{19.11}$$

A rotation matrix is the set of elements of a division algebra that are of unit length, *norm* $=1$. In 2-dimensions, the rotations are seen as being the same in $\mathbb{R}^2$ space as they are in the division algebras. Within higher dimensional spaces, there are great differences between rotations in $\mathbb{R}^n$ spaces and the division algebra spaces. An example is 3-dimensional rotation in the commutative 3-dimensional C_3 division algebra space of the point (a,b,c) given by the rotation matrix:

$$\begin{bmatrix} v_A(\theta,\phi) & v_B(\theta,\phi) & v_C(\theta,\phi) \\ v_C(\theta,\phi) & v_A(\theta,\phi) & v_B(\theta,\phi) \\ v_B(\theta,\phi) & v_C(\theta,\phi) & v_A(\theta,\phi) \end{bmatrix}\begin{bmatrix} a & b & c \\ c & a & b \\ b & c & a \end{bmatrix} \tag{19.12}$$

$v_i(\theta,\phi)$ are the 3-dimensional trigonometric functions. Within the rotation matrix of a *n*-dimensional division algebra space there are *n* different trigonometric functions. The reader's attention is drawn to the presence of two arguments (angles) within the 3-dimensional trigonometric functions, $v_i(\theta,\phi)$. In general, within a *n*-dimensional division algebra space, the trigonometric functions have $(n-1)$

---

[93] Trigonometric functions are projections from a point in the space on to an axis. The are *n* of them in a *n*-dimensional division algebra space – one for each axis. The seemingly obvious fact that an *n*-dimensional space has *n* axes and therefore *n* trigonometric functions has escaped notice by mathematicians for some 1500 years.

arguments (angles) which correspond to the imaginary variables of that algebra in the Cartesian form of the algebra. This multi-argument nature of the rotation matrix functions, the trigonometric functions, is what makes the algebra a spinor algebra.

As with the familiar 2-dimensional rotation matrices, when all arguments (angles) in a higher dimensional rotation matrix are zero, the rotation matrix is the identity. Of course it is; rotation through zero angle is just the identity.

This 3-dimensional space does not have 2-dimensional sub-spaces (2-dimensional planes) [94]. We give an example:

$$
\begin{bmatrix} v_A(\theta,\phi) & v_B(\theta,\phi) & 0 \\ 0 & v_A(\theta,\phi) & v_B(\theta,\phi) \\ v_B(\theta,\phi) & 0 & v_A(\theta,\phi) \end{bmatrix} \begin{bmatrix} a & b & 0 \\ 0 & a & b \\ b & 0 & a \end{bmatrix}
$$

$$
= \begin{bmatrix} \sim & \sim & b.v_B(\theta,\phi) \\ b.v_B(\theta,\phi) & \sim & \sim \\ \sim & b.v_B(\theta,\phi) & \sim \end{bmatrix}
$$

(19.13)

We see that the attempted rotation within a 2-dimensional plane has 'thrown' the point (a,b) out of the assumed 2-dimensional plane.

Since there is no 2-dimensional plane in this space, it is impossible to remain within that non-existent 2-dimensional plane. Within the 3-dimensional C_3 space above, (19.12), we have genuine 3-dimensional rotation and not a combination of 2-dimensional rotations. That is worth emphasising.

> *Emphasis:* Within the 3-dimensional C_3 space above, (19.12), we have genuine 3-dimensional rotation and not a combination of 2-dimensional rotations.

Of course, genuine 3-dimensional rotation is spinor rotation [95].

---

[94] This is because the finite group C_3 does not have a C_2 sub-group.

[95] The 3-dimensional spinors do not exhibit double cover.

Although we cannot rotate in a non-existent 2-dimensional plane in the 3-dimensional division algebra, we can vary only one of the two angles (rotation parameters):

$$
\begin{bmatrix} v_A(\theta,0) & v_B(\theta,0) & v_C(\theta,0) \\ v_C(\theta,0) & v_A(\theta,0) & v_B(\theta,0) \\ v_B(\theta,0) & v_C(\theta,0) & v_A(\theta,0) \end{bmatrix} \begin{bmatrix} a & b & c \\ c & a & b \\ b & c & a \end{bmatrix}
$$
$$
= \begin{bmatrix} av_A(\theta,0) + cv_B(\theta,0) + bv_c(\theta,0) & \sim & \sim \\ \sim & & \sim & \sim \\ \sim & & \sim & \sim \end{bmatrix}
$$

(19.14)

This is a '2-dimensional rotation' in the sense that only one angle varied, but it does not happen in a 2-dimensional plane. In a different co-ordinate system, this '2-dimensional rotation' would involve the changing of both angles, and so we see that it is a 'will-o'the-wisp' '2-dimensional rotation' that is a figment of the choice of co-ordinate system.

Because the higher dimensional trigonometric functions have arguments that correspond to the imaginary variables of the algebra[96], rotations within the higher dimensional division algebra spaces cannot be separated into sets of 2-dimensional rotations in the way they can in the $\mathbb{R}^n$ types of space. This is the central difference between spinor rotation (that's rotation in spinor spaces) and classical rotation (normal rotation to which we are experientially accustomed).

$\mathbb{R}^4$ has six 2-dimensional sub-spaces (planes). Any pairing of two axes will form a 2-dimensional plane. Three of these pairs of axes are purely spatial and the rotation associated with them is of the form of the Euclidean complex plane, $\mathbb{C}$. The 4-dimensional quaternion space has only three 2-dimensional sub-spaces. Only the three pairings of one imaginary element with the identity element form 2-dimensional planes. (These are 4-dimensional 2-dimensional planes – the 4×4 rotation matrix cannot act upon a 2×2 matrix.) These three 2-

---

[96] The angle in the sine and cosine functions is the imaginary variable of the complex plane $\mathbb{C}$.

dimensional planes are associated with three rotations, and the nature of the rotation associated with them is of the form of the Euclidean complex plane, $\mathbb{C}$. This is the famous double cover isomorphism of $SO(3)$ & $SU(2)$.

If we try to rotate in a quaternion 2-dimensional plane formed from a pair of two imaginary axes, we are 'thrown' out of the plane:

$$\begin{bmatrix} 0 & v_BQ & 0 & v_DQ \\ -v_BQ & 0 & -v_DQ & 0 \\ 0 & v_DQ & 0 & -v_BQ \\ -v_DQ & 0 & v_BQ & 0 \end{bmatrix} \begin{bmatrix} 0 & b & 0 & d \\ -b & 0 & -d & 0 \\ 0 & d & 0 & -b \\ -d & 0 & b & 0 \end{bmatrix} \qquad (19.15)$$

$$= \begin{bmatrix} -b.v_BQ - dv_DQ & -b.v_BQ - dv_DQ & -b.v_BQ - dv_DQ & -b.v_BQ - dv_DQ \\ \sim & \sim & \sim & \sim \\ \sim & \sim & \sim & \sim \\ \sim & \sim & \sim & \sim \end{bmatrix}$$

$$(19.16)$$

We see that there are three 2-dimensional $\mathbb{C}$ type rotations in quaternion space, which are spinor rotations, and there are three purely spatial $\mathbb{C}$ type rotations within the 3-dimensional spatial part of our 4-dimensional space-time. You see the resemblance; this is why it is said that $SU(2)$ is isomorphic to $SO(3)$[97].

Within a n-dimensional division algebra, there is always a $n \times n$ rotation matrix (the imaginary part of the polar form of the algebra) containing the n n-dimensional trigonometric functions of the algebra, and, within division algebras, rotation can always be expressed as this single rotation matrix acting upon a point. (In non-commutative division algebras, the rotation matrix is sometimes taken to act twice upon the point, but we will come to that shortly.)

---

[97] Actually, the isomorphism is based upon the commutation relations, but the given resemblance is intuitively more obvious.

Eigenvectors and spinor rotations:

Consider the quaternion rotation matrix (19.1). It has eigenvalues:

$$i\sqrt{b^2+c^2+d^2}, \qquad -i\sqrt{b^2+c^2+d^2}$$
$$i\sqrt{b^2+c^2+d^2}, \qquad -i\sqrt{b^2+c^2+d^2} \qquad (19.17)$$

It has no eigenvectors that are unchanged by rotation (none of the eigenvalues is unity). There is no direction in quaternion space that is unchanged by rotation. There is no 1-dimensional axis about which the rotation happens. The reader should contrast this with the case in $\mathbb{R}^n$ space of which an example is given above, (19.6).

We cannot increase the size of the quaternion matrix by adding an extra row and an extra column with a single 1 in the corner as we increased the size of the $\mathbb{C}$ rotation matrix in (19.8) because the quaternions are a 4-dimensional algebra. It is only because our space-time, formed from the superimposition of the A_3 spaces, is 4-dimensional and has the distance function it does have that we can add extra rows and columns to the 2-dimensional rotations. We can imagine a 5-dimensional space with a quaternion rotation matrix in the top corner and a single 1 in the bottom corner; we would then have a quaternion rotation about the fifth axis, but such a space does not exist; it is certainly not a division algebra space or a $\mathbb{R}^n$ type of space.

A single rotation or a sequence of rotations:

Within $\mathbb{R}^n$ types of space, if I wish to move from starting point A to finishing point B where both points are equally distant from the origin, I must do it using only 2-dimensional rotations. If the co-ordinate system is 'special', I can achieve this in a single 2-dimensional rotation in a single zero plane of the $\mathbb{R}^n$ space, but more generally, I will have to use a sequence of 2-dimensional rotations. An example is a rotation down the Greenwich meridian from London to the equator followed by a rotation along the equator to Nairobi. This is expressed by the fact that we need a different rotation matrix for each 2-dimensional plane in the $\mathbb{R}^n$ space. In general, I rotate by one angle first and then rotate by another angle second etc. in sequence. If I want to combine the two 2-dimensional rotations into a single rotation, then I must, in general, change the co-ordinate system. There is no single rotation matrix that will accomplish all possible rotations in all possible bases.

Simple connection:

In general, the different 2-dimensional rotation planes of $\mathbb{R}^n$ space are each a separate 2-dimensional algebra; they might be the same type of algebra, but they are separate copies of that algebra. Sometimes they are not the same 2-dimensional algebra. Rotation in $\mathbb{R}^n$ space is done by using several disjoint 2-dimensional algebras. Ultimately, this is why rotations in $\mathbb{R}^n$ space, which are identified with the groups $SO(n, p)$, are not simply connected.

Within a division algebra space, because all the angles (rotation parameters) are all within the trigonometric functions in a single rotation matrix, in general, regardless of how the co-ordinate system is aligned, I can move from starting point A to finishing point B in a single rotation. In general, I will have to change every angle in the rotation matrix at the same time. In a division algebra space like the quaternions, see (19.1), it might seem that I can rotate 2-dimensionally by changing one of the angles, I can rotate 3-dimensionally by changing two of the angles, and I can rotate 4-dimensionally by changing all three of the angles, and all within one rotation matrix. Of course, as in the 3-dimensional case, these '2-dimensional rotations' are no more than 'will-o'-the-wisp' products of the choice of co-ordinate system. Because the quaternions are a single algebra, rotation within the quaternions, which is identified with the group $SU(2)$, is simply connected.

Rotations in $\mathbb{R}^n$ space are associated with the $SO(p,q)$ groups; it is well known that these are not simply connected. It is also well known that for each $SO(p,q)$ group there exists a simply connected group which is a double cover of the $SO(p,q)$ group. These double covers are called spin groups, $spin(p,q)$, and the individual elements in them, the individual rotations, are spinors.

This is one, very common, and not foolish, definition of what a (unit length) spinor is; that definition is "a spinor is a rotation element in the simply connected double cover of the $SO(p,q)$ group". It is a sub-definition of our definition of spinors for it picks out only a particular type of spinor.

Summary:
Rotations in division algebra spaces, that is spinor rotations, are different from rotations in our 4-dimensional space-time. Rotations in division algebra spaces are multi-angular rotations in that the n-dimensional trigonometric functions within the single $n \times n$ rotation matrix have n arguments (angles) within them. Rotations in division algebra spaces are not rotations about an axis. Rotation matrices in division algebra spaces are true, honest, proper, and sane rotation matrices.

Rotation in our 4-dimensional space-time are multi-matrix rotations in that they are accomplished by a number of different 'untrue, dishonest, improper, and insane rotation matrices'[98] all of which produce 2-dimensional rotations using two 2-dimensional trigonometric functions which both accept only one argument (angle).

---

[98] Perhaps your author got a little carried away there.

Chapter 20

Commutative Rotations

The rotation group of a geometric space:
Within any geometric space of any number of dimensions, there is a 'spherical surface' of points in that space that are unit distance from the origin. These points form a continuous group (continuous surface) connected to each other by rotation. To every rotation through angle, say (b,c,d), there is an inverse rotation (a conjugate rotation) through angle $(-b,-c,-d)$. There is an identity element which is rotation through zero, $(0,0,0)$, and no rotation can take a point in the 'spherical surface' out of that 'spherical surface'. These 'spherical surfaces' are continuous groups, but we must be careful not to confuse them with classical Lie groups.

Commutative rotations – the notational cacophony:
A rotation matrix acts upon a point (written as a matrix) by matrix multiplication. The rotation matrix can act on the left of the point or on the right of the point. If the division algebra is commutative, then rotation on the left is the same as rotation on the right:

$$\begin{bmatrix} \cos\theta & \sin\theta \\ -\sin\theta & \cos\theta \end{bmatrix} \begin{bmatrix} a & b \\ -b & a \end{bmatrix} = \begin{bmatrix} a & b \\ -b & a \end{bmatrix} \begin{bmatrix} \cos\theta & \sin\theta \\ -\sin\theta & \cos\theta \end{bmatrix} \quad (20.1)$$

Both of these calculations appear 'lopsided'. It would be prettier to act on both left and right of the point. This means multiplying the point by two rotation matrices. If we are to do this, then we must put only half the angle into each matrix to get the same result as with only one rotation matrix.

154

$$\begin{bmatrix} \cos\dfrac{\theta}{2} & \sin\dfrac{\theta}{2} \\ -\sin\dfrac{\theta}{2} & \cos\dfrac{\theta}{2} \end{bmatrix} \begin{bmatrix} a & b \\ -b & a \end{bmatrix} \begin{bmatrix} \cos\dfrac{\theta}{2} & \sin\dfrac{\theta}{2} \\ -\sin\dfrac{\theta}{2} & \cos\dfrac{\theta}{2} \end{bmatrix} \qquad (20.2)$$

In commutative division algebra spaces, a rotation on the right with an equal rotation on the left moves the point through twice the angle:

$$\begin{bmatrix} \cos\theta & \sin\theta \\ -\sin\theta & \cos\theta \end{bmatrix} \begin{bmatrix} a & b \\ -b & a \end{bmatrix} \begin{bmatrix} \cos\theta & \sin\theta \\ -\sin\theta & \cos\theta \end{bmatrix}$$
$$= \begin{bmatrix} \cos\theta & \sin\theta \\ -\sin\theta & \cos\theta \end{bmatrix} \begin{bmatrix} \cos\theta & \sin\theta \\ -\sin\theta & \cos\theta \end{bmatrix} \begin{bmatrix} a & b \\ -b & a \end{bmatrix} \qquad (20.3)$$
$$= \begin{bmatrix} \cos(2\theta) & \sin(2\theta) \\ -\sin(2\theta) & \cos(2\theta) \end{bmatrix} \begin{bmatrix} a & b \\ -b & a \end{bmatrix}$$

We might now pretend that we have to rotate through 720° to get back to where we started, $0^{\circ} = 360^{\circ}$, but we are merely deceiving ourselves. Of course, in commutative spaces, we do not need to balance the rotation matrices and any unequal fractions of θ would achieve the same rotation.

In commutative geometric spaces a rotation on the right with an equal inverse rotation on the left leaves the point unmoved, as does the alternative:

$$\begin{bmatrix} \cos\theta & \sin\theta \\ -\sin\theta & \cos\theta \end{bmatrix}^{-1} \begin{bmatrix} a & b \\ -b & a \end{bmatrix} \begin{bmatrix} \cos\theta & \sin\theta \\ -\sin\theta & \cos\theta \end{bmatrix} = \begin{bmatrix} a & b \\ -b & a \end{bmatrix}$$
$$\begin{bmatrix} \cos\theta & \sin\theta \\ -\sin\theta & \cos\theta \end{bmatrix} \begin{bmatrix} a & b \\ -b & a \end{bmatrix} \begin{bmatrix} \cos\theta & \sin\theta \\ -\sin\theta & \cos\theta \end{bmatrix}^{-1} = \begin{bmatrix} a & b \\ -b & a \end{bmatrix} \qquad (20.4)$$

These are all very familiar to us. We will sometimes refer to rotation by a single rotation matrix as 'single matrix rotation' whereas rotation by two rotation matrices might be referred to as 'double matrix rotation'.

The cacophony quieted:

As we saw above, we can express a rotation in one of several ways, but the complete set of possible rotations (2π 's worth in the case of the Euclidean circle) is the same regardless of the rotation expression we choose. By putting different angles into the rotation matrices, we can rotate to any point in the spherical surface that surrounds the origin. When it is all said and done, provided we stick to one way of applying a rotation matrix or several rotation matrices, we get the full set of possible rotations. All the different ways of writing rotation are no more than notational differences.

Distance functions within division algebra spaces:

Within a *n*-dimensional division algebra, the distance of a point from the origin is simply the nth root of the determinant of that point. Thus, in practice, when dealing with rotation in division algebra spaces (spinor spaces), it is sufficient to consider only the invariance of the determinant[99]. We can then see a rotation matrix within a division algebra as a linear transformation with a determinant of unity because multiplication by that rotation matrix will not change the determinant of the rotated matrix. For example, within the complex plane, $\mathbb{C}$:

$$\det\left(\begin{bmatrix} \cos\theta & \sin\theta \\ -\sin\theta & \cos\theta \end{bmatrix}\right) = \cos^2\theta + \sin^2\theta = 1 \qquad (20.5)$$

Within division algebras, the rotation matrix is formed by taking the exponential of the basic algebraic matrix form with zero real part. This is the exponential of a matrix with zero trace, and so the rotation matrix will always have determinant unity. For example, in three dimensions:

---

[99] Since the determinant of a matrix is the product of the eigenvalues of that matrix, invariance of the determinant (rotation) is invariance of the product of the eigenvalues.

$$\det\left(\exp\begin{bmatrix} 0 & b & c \\ c & 0 & b \\ b & c & 0 \end{bmatrix}\right) = \det\left(\begin{bmatrix} v_A(b,c) & v_B(b,c) & v_C(b,c) \\ v_C(b,c) & v_A(b,c) & v_B(b,c) \\ v_B(b,c) & v_C(b,c) & v_A(b,c) \end{bmatrix}\right) = 1$$

$$(20.6)$$

Any matrix with determinant unity will do:

With thought, we see that, in a *n*-dimensional division algebra space, any linear transformation of the appropriate size (a $n \times n$ matrix) with a determinant of unity is a rotation because multiplication by it will conserve the determinant. Some rotation matrices do not, at first sight, appear to be rotations; for example:

$$\det\left(\begin{bmatrix} \dfrac{1}{2} & \dfrac{\sqrt{3}}{2} \\ -\dfrac{\sqrt{3}}{2} & \dfrac{1}{2} \end{bmatrix}\right) = \frac{1}{4} + \frac{3}{4} = 1 \tag{20.7}$$

or

$$\det\left(\begin{bmatrix} 1.6 & 1.25 \\ 1.25 & 1.6 \end{bmatrix}\right) = 1 \tag{20.8}$$

With deeper insight, we see that the first of these, (20.7), is just a rotation in the complex plane, $\mathbb{C}$, though $\dfrac{\pi}{3}$. This is not an exceptional example. If a matrix has determinant unity, then it is a point that is distance unity from the origin; as such, it is a point in the unit 'sphere' that envelopes the origin; as such, it is equal to a rotation matrix with some values of the angles. However, we must be careful to notice within which space (division algebra) the matrix is a rotation. The second of the above, (20.8), is rotation in space-time through $\dfrac{\pi}{3}$.

Within matrices in general, we have the relationship:

$$\det\left(A^{-1}\right)=\left(\det\left(A\right)\right)^{-1} \qquad (20.9)$$

Thus, using any element of a given division algebra, P, we care to choose, we can write a rotation of the starting point A to the finishing point B as:

$$P^{-1}AP = B \qquad (20.10)$$

Now, the point P need not have determinant unity, and so it need not be on the unit 'sphere', but it is equal to a real number multiplied by a point on the unit 'sphere' because any Cartesian point in a division algebra can be written in its polar form which is a real number multiplied by a rotation matrix. Obviously, (20.9), the determinant of P^{-1} is the inverse of the determinant of P, and so (20.10) conserves the determinant and is a rotation. Writing P in polar form makes this obvious:

$$\left(a[Rot]\right)^{-1} A\left(\frac{1}{a}[Rot]\right)=a\frac{1}{a}[Rot]^{-1} A[Rot]=[Rot]^{-1} A[Rot]= B$$

$$(20.11)$$

The real numbers, $\left\{a,\dfrac{1}{a}\right\}$ commute with everything and cancel.

Division algebra Lie groups:
The set of points given by the n-dimensional ($n\times n$) rotation matrix as the parameters (angles) within it are varied is a $(n-1)$ dimensional 'spherical surface' embedded within the n-dimensional space. An example is the 1-dimensional unit circle embedded in the 2-dimensional complex plane, $\mathbb{C}$; another example is the 3-dimensional set of points (given by the quaternion rotation matrix) that are distance unity from the origin in 4-dimensional quaternion space. Such a 'spherical surface' is a division algebra continuous group; such division algebra continuous groups are commutative within

commutative division algebras as with the unit circle, $U(1)$, within the complex plane, $\mathbb{C}$.

A division algebra continuous group is not necessarily the same thing as a classical Lie group[100] with which Lie algebraists might be more familiar. The division algebra continuous groups exist within division algebra spaces; classical Lie groups exist as 'spherical surfaces' within invented linear spaces. An example of such a difference is the difference between the quaternion rotation matrix and the classical Lie group $SU(2)$ given as the three Pauli matrices; these two groups are isomorphic as Lie groups, but they are not the same thing, and they are expressed differently because they are within different types of space. In general, classical Lie groups do not appear as rotation matrices even though they are presented as being concerned with rotation because they exist in $\{\mathbb{R}^n, \mathbb{C}^n, \mathbb{H}^n, \mathbb{O}^n\}$ types of space.

---

[100] $U(1)$ is exceptional.

Chapter 21

Non-commutative Rotation

We will use the quaternions as an example of a non-commutative division algebra. A quaternion is of the form:

$$\mathbb{H} = Q = \begin{bmatrix} e & f & g & h \\ -f & e & -h & g \\ -g & h & e & -f \\ -h & -g & f & e \end{bmatrix} \tag{21.1}$$

We have called this matrix Q to ease notation later. The quaternion rotation matrix for a rotation through quaternion angle $\{b,c,d\}$ is given by the exponential of the quaternion with zero real variable[101]:

$$\mathbb{H}_{Rot} = \exp\left(\begin{bmatrix} 0 & b & c & d \\ -b & 0 & -d & c \\ -c & d & 0 & -b \\ -d & -c & b & 0 \end{bmatrix}\right) \tag{21.2}$$

$$\mathbb{H}_{Rot} = \begin{bmatrix} \cos(\lambda) & \frac{b}{\lambda}\sin(\lambda) & \frac{c}{\lambda}\sin(\lambda) & \frac{d}{\lambda}\sin(\lambda) \\ -\frac{b}{\lambda}\sin(\lambda) & \cos(\lambda) & -\frac{d}{\lambda}\sin(\lambda) & \frac{c}{\lambda}\sin(\lambda) \\ -\frac{c}{\lambda}\sin(\lambda) & \frac{d}{\lambda}\sin(\lambda) & \cos(\lambda) & -\frac{b}{\lambda}\sin(\lambda) \\ -\frac{d}{\lambda}\sin(\lambda) & -\frac{c}{\lambda}\sin(\lambda) & \frac{b}{\lambda}\sin(\lambda) & \cos(\lambda) \end{bmatrix} \tag{21.3}$$

$$\lambda = \sqrt{b^2 + c^2 + d^2}$$

---

[101] The real variable is the variable on the leading diagonal.

The determinant of the quaternion rotation matrix is unity; of course it is; after all, this is a rotation matrix.

$$\det\left(\mathbb{H}_{Rot}\right)=\left(\cos^2\sqrt{b^2+c^2+d^2}+\sin^2\sqrt{b^2+c^2+d^2}\right)^2=1 \quad (21.4)$$

In general, $\det\left(A^{-1}\right)=\left(\det\left(A\right)\right)^{-1}$, and so the determinant of the inverse of the quaternion rotation matrix is unity, of course.

Note that the quaternion trigonometric functions have three arguments; the 4-dimensional angle is a 3-dimensional vector within the 3-dimensional quaternion 'spherical surface' unit distance from the origin.

If we reverse the variables, $\{b,c,d\}\rightarrow\{-b,-c,-d\}$, that are the quaternion angle, that is we rotate in the reverse direction, that is we take the quaternion conjugate, we get the inverse of the rotation matrix which is also the conjugate of the quaternion rotation matrix:

$$_{REV}\mathbb{H}_{Rot}=\mathbb{H}_{Rot}^{-1}=\mathbb{H}_{Rot}^{*}=$$

$$\begin{bmatrix} \cos(\lambda) & -\dfrac{b}{\lambda}\sin(\lambda) & -\dfrac{c}{\lambda}\sin(\lambda) & -\dfrac{d}{\lambda}\sin(\lambda) \\[2mm] \dfrac{b}{\lambda}\sin(\lambda) & \cos(\lambda) & \dfrac{d}{\lambda}\sin(\lambda) & -\dfrac{c}{\lambda}\sin(\lambda) \\[2mm] \dfrac{c}{\lambda}\sin(\lambda) & -\dfrac{d}{\lambda}\sin(\lambda) & \cos(\lambda) & \dfrac{b}{\lambda}\sin(\lambda) \\[2mm] \dfrac{d}{\lambda}\sin(\lambda) & \dfrac{c}{\lambda}\sin(\lambda) & -\dfrac{b}{\lambda}\sin(\lambda) & \cos(\lambda) \end{bmatrix} \quad (21.5)$$

$$\lambda=\sqrt{b^2+c^2+d^2}$$

The square root, $\lambda=\sqrt{b^2+c^2+d^2}$, can be taken to be either positive or negative; provided we do this consistently, we get the same rotation matrix for both options.

If we did not take the square root sign consistently, we would effectively scatter minus signs about the matrix in a random way and the result would not necessarily be a quaternion, or any other division algebra. We need to maintain the algebraic matrix form to maintain the algebra. The reader is reminded of the S_3 algebra we looked at earlier and in which we discovered the same need for consistency in choice of square root signs needed to maintain the algebra.

We see that the two angles $(b,c,d)\&(-b,-c,-d)$ give the same argument to the $\{\cos(\ ),\sin(\ )\}$ functions and that it is only the presence of the square root outside of the $\{\cos(\ ),\sin(\ )\}$ functions that reverses the direction of the rotation.

Double covers uncovered:

The reader might be aware that in conventional Clifford algebra, within a spin group, we get the same 2-dimensional rotation for both $\{\theta,-\theta\}$ and we call this a double cover - see: (6.1) to (6.6). If we set two of the variables in the quaternion rotation matrix to zero, we have a 4-dimensional quaternion 2-dimensional rotation:

$$\mathbb{H}_{Rot}^{c=d=0} =$$

$$
\begin{bmatrix}
\cos\left(\sqrt{b^2}\right) & \sqrt{+1}\sin\left(\sqrt{b^2}\right) & 0 & 0 \\
-\sqrt{+1}\sin\left(\sqrt{b^2}\right) & \cos\left(\sqrt{b^2}\right) & 0 & 0 \\
0 & 0 & \cos\left(\sqrt{b^2}\right) & -\sqrt{+1}\sin\left(\sqrt{b^2}\right) \\
0 & 0 & \sqrt{+1}\sin\left(\sqrt{b^2}\right) & \cos\left(\sqrt{b^2}\right)
\end{bmatrix}
$$

$$(21.6)$$

We see that we get the same rotation for both the angles $\{b,-b\}$. The quaternions are a double cover of a 2-dimensional rotation because quaternion trigonometric functions have a square root sign in the argument. A 2-dimensional rotation in quaternion space is not a 2-dimensional rotation in 2-dimensional space; the difference is partly the square root sign.

With a little thought, the reader will see that setting two of the variables to zero actually gives two 4-dimensional 2-dimensional quaternion rotation matrices depending upon the sign of the square root outside of the $\sin\left(\sqrt{...}\right)$ functions. These correspond to rotation and reverse rotation.

Defining spinors:
We now come to how we define spinors. If we insist on spinors being a double cover of the orthogonal groups, $SO(p,q)$, then, because double cover results from the square root sign in the trigonometric functions, we have spinors in only the $C_2 \times C_2 \times ...$ algebras. The 3-dimensional rotation matrix of a C_3 algebra does not have this double cover. It is no more than a matter of sophistry whether we refer to all division algebra rotation matrices as spinors or to only rotation matrices of the $C_2 \times C_2 \times ...$ algebras as spinors, but, as we have previously stated, we choose all division algebra rotation matrices.

Left or right rotation:
We now come to a difference between rotation in a commutative division algebra space and rotation in a non-commutative division algebra space.

Multiplying the quaternion, Q, (21.1), on the left by the quaternion rotation matrix, (21.3), gives a result that is different from multiplying the quaternion, Q, (21.1), on the right by the quaternion rotation matrix, (21.3).

So, single matrix rotation through a non-zero given quaternion angle, (b,c,d), will take a point, (f,g,h), to two different other points depending upon whether the single rotation matrix is multiplied on the right or multiplied on the left. If the angle is the same in both cases, the direction of rotation is the same in both cases. So, we rotate in the same direction through the same angle and end up in two different places corresponding to rotation on the right and rotation on the left. The difference in the calculation of these two different points is not a variable but the placement of the single rotation matrix to the left or to the right of the starting point.

Now, within the quaternions, we have these two different single matrix rotations (left or right). We have left rotation of the point A through the angle (b,c,d) to the point B:

$$\mathbb{H}_{Rot_{bcd}} A = B \qquad (21.7)$$

Could we move from the point A to the point B by a rotation on the right if we rotate through a different angle, (e,f,g)? Yes, of course we can because there is only one quaternion 'spherical surface' group. We put:

$$\mathbb{H}_{Rot_{efg}} = A^{-1} \mathbb{H}_{Rot_{bcd}} A$$
$$A\mathbb{H}_{Rot_{efg}} = \mathbb{H}_{Rot_{bcd}} A = B \qquad (21.8)$$

This shows that both left and right rotation move points into the same set of points; this set of points is the 'spherical shell' around the origin in quaternion space. We can move from one point in this shell to any other point in this shell by either left or right rotation if we choose the correct angle. Looking at the first line of (21.8) with (20.10) in mind, we see that rotation on the left differs from rotation on the right by a rotation.

The reader might see ambiguity here. One rotation matrix moves a point to two different destinations. There is no ambiguity provided we consistently apply the rotation matrix to either the left or to the right.

$\mathbb{R}^3$ and the different natures of non-commutativity:

In the $\mathbb{R}^n$ spaces, all rotations are sequences of 2-dimensional rotations. In $\mathbb{R}^3$, (think the surface of the Earth) two such rotations are expressed by matrices like[102]:

$$\begin{bmatrix} \cos\theta_{GM} & \sin\theta_{GM} & 0 \\ -\sin\theta_{GM} & \cos\theta_{GM} & 0 \\ 0 & 0 & 1 \end{bmatrix}, \quad \begin{bmatrix} \cos\phi_{EQ} & 0 & \sin\phi_{EQ} \\ 0 & 1 & 0 \\ -\sin\phi_{EQ} & 0 & \cos\phi_{EQ} \end{bmatrix} \quad (21.9)$$

These matrices are not commutative, and so the two possible orders in which they can be multiplied will give different results. The difference in order is whether we rotate through θ_{GM} first and ϕ_{EQ} second or we rotate through ϕ_{EQ} first and θ_{GM} second. In order to rotate from London to Nairobi in both cases, we need to change (not swap) the angles when we change the order of multiplication. As with the quaternions, two different rotations (opposite order sequences) can both get from A to B if they are through different angles.

In quaternion space, because single matrix rotation is done with a single rotation matrix, we cannot swap the order in which we rotate through each angle. The non-commutativity of $\mathbb{R}^3$ is in the separate rotation matrices. In quaternion space, a point on the quaternion sphere is a quaternion, and this does not commute with the quaternion rotation matrix. We cannot have non-commutativity between different types of rotation matrices within quaternion space because there is only one type of rotation matrix in quaternion space.

We see here a difference in how non-commutativity is manifest within the quaternion space and the $\mathbb{R}^3$ space. In the $\mathbb{R}^3$ space, the rotation angle is split into two 2-dimensional angles; in the quaternion space, there is no such splitting of the angle. The splitting of the angle in $\mathbb{R}^3$ space gives three 3×3 rotation matrices that do not commute with each other. In quaternion space, non-commutativity is intrinsic to the

---

[102] If the reader is curious, the GM subscript relates to Greenwich Meridian, and the EQ subscript relates to Equator.

space. Quaternion rotation is a phenomenon entirely unexperienced by human senses because we live in a space-time that has only 2-dimensional rotation.

Other rotations:

The story continues. Since the determinant of the quaternion rotation matrix is unity and the determinant of the inverse of the quaternion rotation matrix is also unity, we can form double matrix rotations which like:

$$\mathbb{H}_{Rot}\mathcal{Q}\mathbb{H}_{Rot} \quad or \quad \mathbb{H}^{-1}_{Rot}\mathcal{Q}\mathbb{H}^{-1}_{Rot} \tag{21.10}$$

These are rotations because they leave the distance from the origin (the determinant) invariant. Clearly, both of the rotations in (21.10) are of the same form.

Because matrices are associative, and because there is no ambiguity in the placement of the rotation matrix, these rotations, (21.10), will give a unique rotation for a single given angle (same angle in both rotation matrices) rather than two different rotations, left or right, for the given angle. Obviously, reversing the order of the rotation matrices produces the same rotation. So what! Using the quaternion rotation matrix consistently on the left gives a unique rotation for every angle.

Another rotation:

We have a double matrix rotation of the form:

$$\mathbb{H}^{-1}_{Rot}\mathcal{Q}\mathbb{H}_{Rot}$$
$$or \tag{21.11}$$
$$\mathbb{H}_{Rot}\mathcal{Q}\mathbb{H}^{-1}_{Rot}$$

This is the same as (20.10) above except that we have cancelled the real multiple and its inverse. The determinant is certainly invariant under this transformation, and so both (21.11) are a rotation. To preserve the symmetry, we put the same angle into both rotation matrices. In fact, these seemingly two different rotations are the inverses of each other, and so we again have a unique rotation (and its inverse) associated with a single angle. We have:

$$\mathbb{H}_{Rot}^{-1}Q\mathbb{H}_{Rot} = P$$
$$\mathbb{H}_{Rot}P\mathbb{H}_{Rot}^{-1} = \mathbb{H}_{Rot}\mathbb{H}_{Rot}^{-1}Q\mathbb{H}_{Rot}\mathbb{H}_{Rot}^{-1} \quad (21.12)$$
$$= Q$$

All types rotations are operationally the same:
Since $\mathbb{H}_{Rot}Q = B$ is a point in the 'spherical shell' that is invariant distance from the origin, and since $B\mathbb{H}_{Rot}^{-1} = C$ is a point in the same 'spherical shell' that is invariant distance from the origin, then $\mathbb{H}_{Rot}^{-1}Q\mathbb{H}_{Rot}$ is also a point in the same 'spherical shell' that is invariant distance from the origin. This is also true of $\mathbb{H}_{Rot}Q\mathbb{H}_{Rot}^{-1}$. In other words, all the different types of rotation act within the same 'spherical shell'. This is because there is only one such 'spherical shell' for a given distance from the origin. All the types of rotations are able to act on every starting point in this 'spherical shell', and all the types of rotations are able to move that starting point to every possible finishing point in this 'spherical shell'. The only difference between the different types of rotations is the angle they need to move through to get from the starting point to the finishing point.

The story so far:
The rotation $\mathbb{H}_{Rot}Q\mathbb{H}_{Rot}$ gives a unique rotation for a given angle with a unique inverse. The rotation $\mathbb{H}_{Rot}Q\mathbb{H}_{Rot}^{-1}$ does the same. If we want to associate a unique rotation with a given angle in quaternion space, then, unless we are prepared to consistently apply one quaternion rotation matrix always to the same side (left or right), we must accept that rotation through 360° in the rotation matrix takes the starting point through 720°. It seems easier to consistently apply the rotation matrix to the left. We could allow two different rotations to be associated with

a single angle in a single direction, but that offends people who are offended by that sort of thing.

Other possible rotations:

Consider rotations like:

$$\mathbb{H}_{RotA}\mathbb{H}_{RotB}^{-1}...\mathbb{H}_{RotE}^{-1}\mathbb{H}_{RotF}^{-1}Q\mathbb{H}_{RotG}\mathbb{H}_{RotH}...\mathbb{H}_{RotJ}\mathbb{H}_{RotM}^{-1}...\mathbb{H}_{RotN}^{-1} \quad (21.13)$$

We see that, since two rotations multiplied together are a rotation, and matrices are associative, rotations such as these must eventually reduce to one of the types of rotations already considered above. Simply reversing the order of any two of the rotation matrices will give a different rotation, but, none-the-less, this will still be a rotation of a type already considered.

Summary of quaternion rotations:

Because quaternions are non-commutative, there are two rotations associated with each angle; these are left rotation and right rotation. There are also two angles associated with moving from a given starting point, A, to a given finishing point, B. Conventionally, only by 'double rotating', that is multiplying on both the left and the right at the same time, can we have a unique rotation associated with a given angle and a unique angle associated with movement from a point A to a point B.

However, if we were to consistently choose to use a single rotation matrix and consistently use this rotation matrix of the left, we would then have a unique angle associated with each rotation and the rotation would cover the whole of the points in the 'spherical shell'. Perhaps there is no need of any more than single matrix left multiplication.

Chapter 22

Rotation and Spinors Again

The reader might be aware that there are many definitions of spinors within Clifford algebra and elsewhere; in this chapter, we first tidy a few points, and then we look at some of the different definitions of spinors which the reader might encounter.

Intrinsic spin:

Physicists tell us that there are two types of angular momentum, but, it seems, there is only one type of mass. These two types of angular momentum are called orbital angular momentum and intrinsic spin. Of these, we are experientially familiar with only the orbital type of angular momentum. We have seen above that there are two types of rotation, the quaternion type of rotation in division algebra space and the sequence of 2-dimensional rotations type of rotation in $\mathbb{R}^n$ types of space. Since we have two types of rotation in the universe, it ought not to surprise us that we have two types of angular momentum. We associate the familiar orbital angular momentum with rotation in the three spatial dimensions of our 4-dimensional space-time. We are therefore left to associate intrinsic spin with the type of rotation in division algebra space.

Spin groups:

Clifford algebraists associate spin groups with unit length spinors, and so division algebra space is what might otherwise be called spinor space. Rather than consider every element of a division algebra (say every quaternion) to be a spinor, it is easier to use the term spinor to mean only the unit length elements of a division algebra, and so, the set of elements that are the 'spherical surface' (the rotation matrix) are the set of spinors. Any other element of the division algebra is just a spinor multiplied by a real number (think polar form of the algebra). Clearly, different division algebras have different spinors and different types of spinors.

But I thought a spinor was two complex numbers:

Within our 4-dimensional space-time, we cannot see 4-dimensional rotations like a quaternion rotation. How would a quaternion rotation (intrinsic spin) look to a person trapped in 4-dimensional space-time who can see only 2-dimensional rotations. We see only rotations in the 2-dimensional division algebra spaces. Presumably, an observer in space-time would see the four variables of the 4-dimensional quaternion rotation as two 2-dimensional rotations.

How would a quaternion appear to a person in 4-dimensional space-time rather than in quaternion space. A quaternion has four independent variables. It seems that a quaternion is split into two complex numbers when it appears in space-time.

Traditionally, $SU(2)$ is presented as a unit length quaternion in a 2×2 matrix containing two complex numbers:

$$\begin{bmatrix} Z_1 & Z_2 \\ -Z_2^* & Z_1^* \end{bmatrix} = \begin{bmatrix} a+ib & c+id \\ -c+id & a-ib \end{bmatrix} \equiv \begin{bmatrix} \begin{bmatrix} a & b \\ -b & a \end{bmatrix} & \begin{bmatrix} c & d \\ -d & c \end{bmatrix} \\ \begin{bmatrix} -c & d \\ -d & -c \end{bmatrix} & \begin{bmatrix} a & -b \\ b & a \end{bmatrix} \end{bmatrix} \in \mathbb{H}$$

(22.1)

It seems that tradition has already split the quaternion into a pair of complex numbers for us.

Clearly, if spinors are elements of the rotation matrix of a division algebra, then an element of the 2-dimensional rotation matrix in the complex plane is a spinor. This is often accepted to be the case in conventional Clifford algebra[103]. There will also be spinors that are elements of the rotation matrix of higher dimensional division algebras such as the $C_2 \times C_2 \times C_2$ division algebras which contain the 8-dimensional Clifford algebras.

Other definitions of spinors:
The word spinor was coined by Paul Ehrenfest. There is no shortage of definitions of spinors.

Spinors are n-tuples of complex numbers:
We have mentioned this view above. A 2-dimensional spinor is seen as a two-component column matrix with entries in $\mathbb{C}$ – a pair of complex numbers[104]. A 4-dimensional spinor is seen as a 4-tuple of complex numbers. Thus, spinors are vectors in $\mathbb{C}^n$ space. This is the view that a general spinor is an n-tuple of complex numbers, but the story is more complicated than just this and not any n-tuple will do. Examples of this view are the two complex numbers that are the spinor we find as the wave-function in the Pauli Schrödinger equation and the four complex numbers that are the Dirac spinor in the Dirac equation:

$$\psi = \begin{bmatrix} a+ib \\ c+id \end{bmatrix} \quad or \quad \begin{bmatrix} a+ib \\ c+id \\ e+if \\ g+ih \end{bmatrix} \qquad (22.2)$$

In your author's opinion, this view is in error because $\mathbb{C}^n$ space does not exist for $n > 1$.

Our rejection of this view means that we must rewrite the Dirac equation and the Schrödinger equation.

---

[103] Clifford algebraists are a little ambiguous about this.
[104] Lounesto: page 52.

Spinors are minimal left ideals of Clifford algebras:[105]

The essential concept here is that the set of spinors is a sub-space of the Clifford algebra. *"The spinors form a sub-algebra of geometric algebra."*[106] We know that sub-algebras of Clifford algebras are themselves Clifford algebras, and so, if the spinor sub-space is also a sub-algebra, we have the view that an entire Clifford algebra is a spinor space. In a slightly different view considered later in this list, the spinor sub-space is identified with a sub-algebra, and so we are on the way to seeing Clifford algebras (and, in our rewrite division algebras) as spinor algebras.

In this view, advantage is taken of the fact that Clifford algebras are isomorphic to matrix algebras[107]. We demonstrate with $Cl_{3,0}$:

$$Cl_{3,0} \cong Mat(2, \mathbb{C}) \tag{22.3}$$

Starting with the 'spinors are n-tuples of complex numbers' view, we write the spinor as a 2×2 matrix:

$$\begin{bmatrix} a+ib \\ c+id \end{bmatrix} \equiv \begin{bmatrix} a+ib & 0 \\ c+id & 0 \end{bmatrix} \tag{22.4}$$

We then use the 2×2 matrix[108]:

$$f = \begin{bmatrix} 1 & 0 \\ 0 & 0 \end{bmatrix} \tag{22.5}$$

to select out a sub-space of the matrix algebra that is a minimal left ideal of the Clifford algebra[109]:

---

[105] Lounesto: page 60.

[106] David Hestenes: Vectors, Spinors, and Complex Numbers in Classical and Quantum Physics: American Journal of Physics Vol 39/9 1013-1027 September 1971 - page 14.

[107] G. Juvet. Operateurs de Dirac et equations de Maxwell
[108] Lounesto: page 60
[109] Marcel Riesz 1947.

$$\psi = S_{Left-Ideal} = \begin{bmatrix} a+ib & 0 \\ c+id & 0 \end{bmatrix} \begin{bmatrix} 1 & 0 \\ 0 & 0 \end{bmatrix} \in Cl_{3,0} f \qquad (22.6)$$

Since spinors are a sub-space (ideal) of $Mat(2,\mathbb{C}) \cong Cl_{3,0}$, they have a basis. In $Mat(2,\mathbb{C})$, that sub-space basis is:

$$f_0 = \begin{bmatrix} 1 & 0 \\ 0 & 0 \end{bmatrix}, \quad f_1 = \begin{bmatrix} 0 & 0 \\ i & 0 \end{bmatrix}, \quad f_2 = \begin{bmatrix} 0 & 0 \\ -1 & 0 \end{bmatrix}, \quad f_3 = \begin{bmatrix} i & 0 \\ 0 & 0 \end{bmatrix} \qquad (22.7)$$

This is equivalent in $Cl_{3,0}$, to[110]:

$$f_0 = \frac{1}{2}\left(1+\vec{e_3}\right), \quad f_1 = \frac{1}{2}\left(\vec{e_{23}}+\vec{e_2}\right)$$
$$f_2 = \frac{1}{2}\left(\vec{e_{31}}-\vec{e_1}\right), \quad f_3 = \frac{1}{2}\left(\vec{e_{12}}+\vec{e_{123}}\right) \qquad (22.8)$$

With a little manipulation, we convert this subspace of the Clifford algebra into spinor space. The minimal left ideal (sub-space) is often identified with the even sub-algebra of the particular Clifford algebra.

We are not restricted to only two component spinors. Dirac spinors can be seen as elements of a minimal left ideal in $Cl_{1,3}$. In the Dirac case, the matrices are 4×4 matrices.[111]

Spinors certainly occur in Clifford algebras, but, it seems to your author, this view is a basically the same as the 'spinors are *n*-tuples of complex numbers' view introduced into Clifford algebras by sleight of notation.

---

[110] Lounesto: page 60
[111] Lounesto: pages 138, 139

Spinors are the something to do with even sub-algebras of Clifford algebras:[112],[113]

Thus, the quaternions, being the even sub-algebra of $Cl_{3,0}$, are something to do with a spinor. *"... the word spinor will be used hereafter to designate an even geometric."*[114]. Of course, a sub-algebra is also a sub-space. In the view of many authors, quaternions are one type of spinors. In the view expressed above, a two component spinor would be two complex numbers. In this even sub-algebra view, the four independent variables of the two complex numbers are now the four variables of a quaternion. The even sub-algebra of $Cl_{3,0}$ is represented by the matrices:

$$Cl_{3,0}^{+} = \begin{bmatrix} w+iz & y+ix \\ -y+ix & w-iz \end{bmatrix} \equiv \begin{bmatrix} w & z & y & x \\ -z & w & -x & y \\ -y & x & w & -z \\ -x & -y & z & w \end{bmatrix} \in \mathbb{H} \qquad (22.9)$$

We also have:

$$U(1) = \{\mathbb{C} : zz^* = 1\} = Cl_{2,0}^{+} \qquad (22.10)$$

This even sub-algebra view alone is too crude. We know that spinors are representations of rotations, and so a spinor might be the rotation matrix of an even sub-algebra of a Clifford algebra, but, to have such a rotation matrix, the Clifford algebra must be presented as a division algebra. We also have that $Cl_{0,2} \cong \mathbb{H}$. Historically, Clifford algebraists have been unaware of division algebras other than $\{\mathbb{C}, \mathbb{H}\}$, and so they have ignored the other division algebras like $\{\mathbb{S}, A_3, ...\}$. It would be consistent to treat all sub-algebras equally rather than favour only the even sub-algebras. We have seen that when a Clifford algebra is

---

[112] David Hestenes: Space-time algebra & David Hestenes: Real Spinor Fields

[113] Lounesto: page 63

[114] David Hestenes: Vectors, Spinors, and Complex Numbers in Classical and Quantum Physics: American Journal of Physics Vol 39/9 1013-1027 September 1971 - page 13.

written as a $C_2 \times C_2 \times ...$ division algebra, the distinction between even elements and odd elements is extinguished. We are thus compelled towards the view that '*spinors are something to do with the sub-algebras of Clifford algebras*'. In fact, all sub-algebras of Clifford algebras are themselves Clifford algebras[115], and so we can forget all about sub-algebras and just take the full division algebra. This is what we have done above in the definition we have chosen.

Spinors are representations of $SO(n)$ Lie groups that cannot be formed by the usual tensor constructions:

There are some representations of $SO(n)$ Lie groups that cannot be formed by the usual tensor constructions. Within Clifford algebra, such representations are called spin representations. The constituents of a spin representation are called spinors. A spinor must be an element of a representation of the double cover of a $SO(n, \mathbb{R})$ with distance function with signature (p,q). These double covers are Lie groups called spin groups $spin(n)$ or $spin(p,q)$.

For example, the mantra is that the spin group $spin(2)$ is a double cover of $SO(2)$ because there are two elements of $spin(2)$ for each element of $SO(2)$. $spin(1)$ is isomorphic to the two points $\{-1, +1\}$, and $spin(2)$ is isomorphic to a circle. Your author is unconvinced of the credence of the mantra.

The spin groups are written as:

$$spin(3) \cong SU(2) \cong Sp(1) \qquad (22.11)$$

---

[115] We allow that the 2-dimensional commutative C_2 algebras and the other commutative algebras like $\{A_1, A_2\}$ that derive from the $C_2 \times C_2 \times ...$ groups are Clifford algebras.

$$spin(5) \cong Sp(2)$$
$$spin(2k+1) \not\cong Sp(k) \ \forall \ k \geq 3 \tag{22.12}$$

Wherein we have also pointed out this particular isomorphism. Basically, rotations in particular types of algebras are written as:

$$\{R \in Mat(N, \mathbb{R}) : R^T R = I, \det(R) = 1\} \sim SO(N)$$
$$\{U \in Mat(N, \mathbb{C}) : U^T U = I, \det(U) = 1\} \sim SU(N) \tag{22.13}$$
$$\{s \in Cl^+_{n,p} : s\bar{s} = 1\} \sim spin(n, p)$$

The spin groups are easily constructed for $\mathbb{R}^{n,p}$ as the set of elements of the Clifford algebra $Cl_{n,p}$ of unit length.[116]

$$spin(n, p) = \{s \in Cl_{n,p} : \tilde{s}s = 1, \bar{s}s = 1\} \tag{22.14}$$

They rotate a vector in $\mathbb{R}^{n,p}$ as:

$$\mathbb{R}^3 \to \mathbb{R}^3 \ : \ \vec{x} \to s\vec{x}s^{-1} \tag{22.15}$$

From our point of view, it is hardly surprising that spinors cannot be deduced from tensors because tensors higher than rank one do not exist within a division algebra. We address the classical Lie groups later in this chapter.

Spinors were defined by Elie Cartan[117] as 2-dimensional complex vectors that represent 3-dimensional complex vectors whose dot-product with themselves is zero:

$$\begin{bmatrix} a+ib \\ c+id \\ e+if \end{bmatrix} \cdot \begin{bmatrix} a+ib \\ c+id \\ e+if \end{bmatrix} = a^2 - b^2 + c^2 - d^2 + e^2 - f^2 + 2i(ab + cd + ef) = 0$$

$$\tag{22.16}$$

---

[116] Lounesto: page 59.
[117] The Theory of Spinors: page 41: Elie Cartan. ISBN: 0-486-64-070-1 1966

Complex vectors whose dot-product with themselves is zero are said to be isotropic. It can be shown that the set of isotropic $\mathbb{C}^3$ vectors (6 independent variables) is a $\mathbb{C}^2$ surface (4 independent variables) parameterized by two complex co-ordinates (Z_0, Z_1) such that:

$$Z_0 = \pm\sqrt{\frac{a+ib-i(c+id)}{2}}, \quad Z_1 = \pm\sqrt{\frac{-a-ib-i(c+id)}{2}} \quad (22.17)$$

This complex 2-dimensional surface is a representation of the complex 3-dimensional isotropic vector given by:

$$a+ib = Z_0^2 - Z_1^2 \quad = \frac{a+ib-i(c+id)}{2} + \frac{a+ib+i(c+id)}{2}$$

$$c+id = i(Z_0^2 + Z_1^2) \quad = i\left(\frac{a+ib-i(c+id)}{2} - \frac{a+ib+i(c+id)}{2}\right)$$

$$e+if = -2Z_0 Z_1 \quad = -2\sqrt{\frac{a+ib-i(c+id)}{2}}\sqrt{\frac{-a-ib-i(c+id)}{2}}$$

$$(22.18)$$

The pair of complex numbers, $Z_0 \,\&\, Z_1$ is a spinor.

Looking at the above, we see that there are two vectors in the $\mathbb{C}^2$ surface that both give the same $\mathbb{C}^3$ vector, $(Z_0, Z_1) \,\&\, (-Z_0, -Z_1)$. Within the complex 2-dimensional surface, there are two points corresponding to the single point in in the complex 3-dimensional space. The two complex 2-dimensional vectors are the two spinors of the complex 3-dimensional isotropic vector. The set of these isotropic vectors is the $\mathbb{C}^2$ spinor sub-space of the $\mathbb{C}^3$ space; each element ($\mathbb{C}^2$ vector) of this sub-space is a spinor.

We can think of a quaternion as being a composit of three 2-dimensional complex numbers, one for each imaginary variable. Cartan is saying that this composite can be associated with two complex numbers. This is the association between the two component spinor of the physicist and a quaternion. It is not a pretty association, but it is an association.

Let us rotate (2-dimensionally as we are accustomed) through the angle θ within the 3-dimensional complex space:

$$\begin{bmatrix} a+ib \\ c+id \\ e+if \end{bmatrix} \rightarrow e^{-i\theta} \begin{bmatrix} a+ib \\ c+id \\ e+if \end{bmatrix} \qquad (22.19)$$

(The minus sign is arbitrary; we could equally well use $e^{i\alpha}$.) We have:

$$Z_0 = \pm\sqrt{\frac{e^{-i\alpha}\left(a+ib-i(c+id)\right)}{2}} = \pm e^{-i\frac{\alpha}{2}}\sqrt{\frac{a+ib-i(c+id)}{2}}$$

$$Z_1 = \pm\sqrt{\frac{e^{-i\alpha}\left(-a-ib-i(c+id)\right)}{2}} = \pm e^{-i\frac{\alpha}{2}}\sqrt{\frac{-a-ib-i(c+id)}{2}}$$

$$(22.20)$$

When the isotropic $\mathbb{C}^3$ vector is rotated by 2π the $\mathbb{C}^2$ spinor is rotated by only π. We can think of a spinor as a *"polarised"*[118] isotropic vector.

Of course, we have the view that $\mathbb{C}^3$ space does not exist. None-the-less, your author is not sure what to make of this definition. It is certainly outdated and has been superseded by more recent definitions within Clifford algebras.

Spinors are modules over Clifford algebras:
This is a view of spinors that is a development by Atiyah, Bott, and Shapiro in 1964 of the minimal left ideal view with which we dealt above. They took this view because it allows differentiation of spinor valued functions on manifolds rather than on only flat spaces. Your author has the view that one can differentiate spinors within only the division algebra of which they are an element and that a manifold is formed by the superimposition of isomorphic division algebras. Your

---

[118] Cartan page 42

author is therefore of the view that differentiation of spinors over manifolds is questionable.

Spinors are groups of invertible elements within Clifford algebras:[119]
This is basically saying that spinors are a division algebra inside of a Clifford algebra. It is not unusual to specifically take the unit length quaternions to be the group $spin(3)$. Since we are rewriting Clifford algebras as division algebras, all sub-algebras are groups of invertible elements including the whole algebra. In this view, the whole sub-algebra is the set of spinors; we prefer the view that only the unit length elements of the algebra, the rotation matrix in polar form, are spinors. *"...unitary spinors are numbers which directly represent rotations..."*[120]

Spinors are elements of the compact spin groups, $Spin(n)$, and are exponentials of bivectors:[121]
The rotation matrix of a division algebra is just the matrix exponential of the imaginary variables of an algebra. We include all the imaginary variables at once, and so we get a multi-angle rotation matrix. If we took the exponential of a single imaginary variable in a $C_2 \times C_2 \times...$ algebra, because every imaginary variable in these algebras forms a 2-dimensional sub-algebra, we would effectively get a 2-dimensional rotation matrix. We see that this 'exponential of bi-vectors' view of spinors is a view that separates the division algebra rotation matrix into separate 2-dimensional rotation matrices. We see a confusion here between multi-angular rotations and sets of 2-dimensional rotations.

---

[119] Topics in Representation Theory: Spin Groups
[120] David Hestenes: Vectors, Spinors, and Complex Numbers in Classical and Quantum Physics: American Journal of Physics Vol 39/9 1013-1027 September 1971 - page 16.
[121] Lounesto: Page 223

Spin groups are the exponentials of the quadratic elements of a Clifford algebra:[122]

The quadratic elements are the even elements of the Clifford algebra. Taking the exponential of these produces a division algebra rotation matrix. This definition coincides, not very clearly but accurately, with the definition of a spin group as the rotation matrix of a division algebra which we propose. Of course, we take the exponential of the whole Clifford algebra, and so the obsession with the even elements is not needed. This obsession derives from the lack of knowledge of the division algebras other than the quaternions and the complex numbers. *"The most succinct expression for a spinor field is:*

$$\psi(x) = e^{\frac{1}{2}\mu(x)} \tag{22.21}$$

Where $\mu(x)$ is an even d-number (Dirac –number).[123]

Spinors are even geometrics:

We follow David Hestines[124] and Pertti Lounesto[125]. Hestenes takes spinors to be 'even geometrics'. By this he means a spinor is an element of the even sub-algebra of a Clifford algebra. The even sub-algebra of a Clifford algebra is based upon the set of all basis elements that are the product of an even number of basis vectors. In $Cl_{2,0}$, there are four basis elements:

$$1, \quad \overrightarrow{e_1}, \quad \overrightarrow{e_1}, \quad \overrightarrow{e_{12}} \tag{22.22}$$

Of these four basis elements, the scalar, 1, and the bi-vector $\overrightarrow{e_{12}}$ are products of an even number of basis vectors. The even sub-algebra is

---

[122] Topics in Representation Theory: Spin groups

[123] David Hestenes: Real Spinor Fields. Journal of Mathematical Physics 8 No. 4 (1967) 798-808: Pages 7-8

[124] David Hestines. Vectors, Spinors, and Complex numbers in Classical and Quantum Physics. American Journal of Physics Vol: 39/9 1013-1027 September 1971.

[125] Pertti Lounesto

therefore based on the two elements $\left\{1,\ \overrightarrow{e_{12}}\right\}$. An element of this sub-algebra is of the form $a+b\overrightarrow{e_{12}}$. This sub-algebra is isomorphic to the complex numbers, $\mathbb{C}$, and so a single complex number is a spinor:

$$spinor = a+b\overrightarrow{e_{12}} = a+ib \tag{22.23}$$

The even sub-algebra of $Cl_{3,0}$ is based on the elements $\left\{1,\ \overrightarrow{e_{12}},\ \overrightarrow{e_{13}},\ \overrightarrow{e_{23}}\right\}$. This sub-algebra is isomorphic to the quaternions; *"A quaternion is a spinor"*[126]:

$$spinor = a+b\overrightarrow{e_{12}}+c\overrightarrow{e_{13}}+d\overrightarrow{e_{23}} = a+\hat{i}b+jc+kd \tag{22.24}$$

Obviously, we have two different spinors here. As we said above, we take the view that we cannot separate out only the even sub-algebras but must consider all sub-algebras.

Pauli matrices:
The Pauli matrices are sometimes used to construct a spinor. The Pauli matrices are:

$$\sigma_1 = \begin{bmatrix} 0 & 1 \\ 1 & 0 \end{bmatrix}, \quad \sigma_2 = \begin{bmatrix} 0 & -i \\ i & 0 \end{bmatrix}, \quad \sigma_3 = \begin{bmatrix} 1 & 0 \\ 0 & -1 \end{bmatrix} \tag{22.25}$$

We have:

$$\begin{aligned} (\sigma_1)^2 = (\sigma_2)^2 = (\sigma_3)^2 &= 1 \\ \sigma_1\sigma_2 &= -\sigma_2\sigma_1 \\ \sigma_1\sigma_3 &= -\sigma_3\sigma_1 \\ \sigma_2\sigma_3 &= -\sigma_3\sigma_2 \end{aligned} \tag{22.26}$$

---

[126] David Hestenes: Vectors, Spinors, and Complex Numbers in Classical and Quantum Physics: American Journal of Physics Vol 39/9 1013-1027 September 1971 - page 14.

We see that the Pauli matrices are the basis vectors of the $Cl_{3,0}$ algebra[127]. Adding in the identity[128], we have the basis elements of $Cl_{3,0}$:

$$
\begin{array}{cc}
I & 1 \\[4pt]
\sigma_1,\ \sigma_2,\ \sigma_3 & \vec{e}_1,\ \vec{e}_2,\ \vec{e}_3 \\[4pt]
\sigma_1\sigma_2,\ \sigma_1\sigma_3,\ \sigma_2\sigma_3 & \vec{e}_{12},\ \vec{e}_{13},\ \vec{e}_{23} \\[4pt]
\sigma_1\sigma_2\sigma_3 & \vec{e}_{123}
\end{array}
\tag{22.27}
$$

We have:

$$
\sigma_{12} = \begin{bmatrix} 0 & 1 \\ 1 & 0 \end{bmatrix}\begin{bmatrix} 0 & -i \\ i & 0 \end{bmatrix} = \begin{bmatrix} i & 0 \\ 0 & -i \end{bmatrix}
\tag{22.28}
$$

$$
\sigma_{13} = \begin{bmatrix} 0 & 1 \\ 1 & 0 \end{bmatrix}\begin{bmatrix} 1 & 0 \\ 0 & -1 \end{bmatrix} = \begin{bmatrix} 0 & -1 \\ 1 & 0 \end{bmatrix}
\tag{22.29}
$$

$$
\sigma_{23} = \begin{bmatrix} 0 & -i \\ i & 0 \end{bmatrix}\begin{bmatrix} 1 & 0 \\ 0 & -1 \end{bmatrix} = \begin{bmatrix} 0 & i \\ i & 0 \end{bmatrix}
\tag{22.30}
$$

Using the Pauli matrices equivalences, we have:

$$
w + x\vec{e}_{23} + y\vec{e}_{31} + z\vec{e}_{12} \equiv \begin{bmatrix} w+iz & y+ix \\ -y+ix & w-iz \end{bmatrix} \equiv \begin{bmatrix} w & z & y & x \\ -z & w & -x & y \\ -y & x & w & -z \\ -x & -y & z & w \end{bmatrix} \in \mathbb{H}
\tag{22.31}
$$

We see that the even sub-algebra of $Cl_{3,0}$ is the quaternions. Thus, following Hestenes, we can see a spinor as a sum of a real number and all three products of two Pauli matrices.

---

[127] Lounesto page 54
[128] Technically, we do not have to add the identity because it arises as a square of a Pauli matrix, but it is not usually included as one of the Pauli matrices.

A conventional list of spinors with comments:

We are grateful to the Wikipedia article entitled spinor for the below listing.

In 1-dimension, a spinor is trivially a real number.

Comment: The real numbers are the division algebra that derives from the finite group C_1; this is only the identity.

In two Euclidean dimensions, left-handed and right-handed Weyl spinors are complex numbers rotated respectively clockwise or anti-clockwise in the complex plane.

In three Euclidean dimensions, the existence of spinors follows from the isomorphism of $SO(3) \& SU(2)$. The spinors are quaternions.

Comment: n-dimension Euclidean space is a space whose distance function is the sum (all positive signature) of *n* squared variables. There is no 3-dimensional geometric space that is Euclidean. Quaternion spinors are associated with the 3-dimensional spatial part of our 4-dimensional space-time because there are three 2-dimensional rotations in quaternion space corresponding to the three $\mathbb{C}$ sub-algebras.

In four Euclidean dimensions, the isomorphism $spin(4) \cong SU(2) \times SU(2)$ leads to two Weyl spinors which are each composed of two quaternion components.

Comment: There is no 4-dimensional geometric space that is Euclidean. There is a $C_2 \times C_2 \times C_2$ 8-dimensional division algebra that is effectively two quaternion sub-algebras 'tied' together.

In higher dimensions, there are no Euclidean geometric spaces, and so we are of the view that higher dimensional spinors in such spaces are fictional. If we wished, we could just follow the Clifford algebras into higher dimensions. Of course, there are spinors in all division algebras of all dimensions, but they do not coincide other than coincidently, if at all, with a conventional listing in higher dimensions.

A different presentation of spinor rotation:

Within Wikipedia, there is an article entitled 'spinor'. Within that article, we have:

"The even graded element

$$\gamma = \frac{1}{\sqrt{2}}\left(1 - \sigma_1\sigma_2\right) \tag{22.32}$$

Corresponds to a vector rotation of 90° around towards σ_2, which can be checked by confirming that:

$$\frac{1}{2}\left(1 - \sigma_1\sigma_2\right)\left(a_1\sigma_1 + a_2\sigma_2\right)\left(1 - \sigma_1\sigma_2\right) = a_1\sigma_2 - a_2\sigma_1 \tag{22.33}$$

It corresponds to a spinor rotation of only 45°, however:

$$\frac{1}{\sqrt{2}}\left(1 - \sigma_1\sigma_2\right)\left(a_1\sigma_1 + a_2\sigma_2\right) = \frac{a_1 + a_2}{\sqrt{2}} + \frac{-a_1 + a_2}{\sqrt{2}}\sigma_1\sigma_2 \tag{22.34}"$$

From this, it might seem that the only difference between a vector rotation and a spinor rotation is the number of times we apply the rotation operator. While being both grateful and respectful of Wikipedia, your author opines that this is not a correct interpretation.

What a tangle:

Reading through the above, the reader will doubtless feel a little exasperated at the seemingly disparate proliferation of definitions of a spinor. Such a proliferation of definitions is indicative of a confusion within modern mathematics over the nature of spinors. One is wary of adding to the confusion by introducing another definition, but we think the definition we have given at the beginning of this book is simple and correct. Within mathematics, simplicity is next to godliness.

The classical Lie groups:

"A useful way of thinking about the classical Lie groups is as rotations in various spaces." – Howard Georgi[129].

The classical Lie groups are sets of 2-dimensional rotations in linear spaces. There is not a single higher dimensional rotation to be found in the whole of Lie group theory. The classical Lie groups are very different from division algebra rotations; they are similar to $\mathbb{R}^n$ rotations.

The classical Lie groups are classified into four families (of infinite number) with five exceptional groups. The four families are the special orthogonal groups of even dimension $D_n = SO(2n)$, the special orthogonal groups of odd dimension $B_n = SO(2n+1)$, the special unitary groups $A_n = SU(n+1)$, and the symplectic groups $C_n = Sp(2n)$. The five exception Lie groups are $\{G_2, F_4, E_6, E_7, E_8\}$.

The $SO(2n)$ groups are rotations[130] in $\mathbb{R}^{2n}$. The $SO(2n+1)$ groups are rotations[131] in $\mathbb{R}^{2n+1}$. The $SU(n+1)$ groups are rotations in $\mathbb{C}^{n+1}$. The $Sp(2n)$ groups are rotations[132] in $\mathbb{H}^{2n}$. The exceptional groups are connected to the octonians, but, since the octonians are non-associative, they cannot form groups[133]. Classical Lie groups are formed by associating a single parameter (single angle) with each element of a group[134]. In this way, all classical Lie group rotations are formed as sets of 2-dimensional rotations.

Some of the classical Lie groups coincide 'accidently' with rotations in the types of space we have previously mentioned;

---

[129] Georgi: Page 240
[130] Georgi: Page 237
[131] Georgi: Page 238
[132] Georgi: Page 241
[133] The octonians are not a division algebra and so do not have a polar form and so do not have a rotation matrix that is a 'spherical surface' group.
[134] Georgi: Page 43

$SU(2)$ & $SO(3,1)$ are examples. Most of the classical Lie groups are associated with spaces which do not exist as geometric spaces derived from finite groups or by superimposition but exist as only linear spaces invented by mathematicians. The classical Lie groups preserve the inner product that is imposed upon these linear spaces by mathematicians, and in that sense they are rotations within the invented linear spaces. These inner products are quadratic forms, usually, but not always, of all positive signature, like:

$$d^2 = a^2 + b^2 + c^2 + d^2 + e^2 + ...$$ (22.35)

Such distance functions include within them the distance function associated with the 2-dimensional $\{\cos(\ ), \sin(\ )\}$ functions, and so 2-dimensional euclidean rotations are associated with these linear spaces. Since quantum mechanical systems are written as linear spaces, rotation within these linear spaces leaves the system invariant and we call this a symmetry of the system. Of course, the only type of rotations we can observe within our 4-dimensional space-time are the 2-dimensional rotations.

Except in the cases of the two 2-dimensional division algebras and the eight non-commutative 4-dimensional division algebras, there are no geometric spaces with distance functions with a quadratic form like (22.35), and so it might seem that the Lie algebras are nothing to do with physics and are no more than figments of the imaginations of mathematicians. This is not quite so. The special orthogonal group $SO(3)$ is connected to spatial rotations in our 4-dimensional space-time and hence connected to orbital angular momentum. The special orthogonal Lie groups, $SO(n)$, have spinor representations, and, within our 4-dimensional space-time, there is something like a 3-dimensional sub-space[135]. Rotations in this 3-dimensional sub-space are not the spinor rotations of a division algebra but are the Lie group $SO(3)$ - which is really a set of 2-dimensional rotations. Further,

---

[135] It is only 'something like' a sub-space because we can never be rid of time.

sitting in our space-time, it does seem that we perceive a distorted form of the division algebras as linear spaces.

Representations of Lie groups:
The classical Lie groups can be represented in different dimensions. In n-dimensional linear space, the Lie groups are represented by $n \times n$ matrices. We thus have the $SU(2)$ type of rotation in 2-dimensional space represented as three 2×2 matrices and we have the $SU(2)$ type of rotation in 3-dimensional space represented as three 3×3 matrices and we have the $SU(2)$ type of rotation in 4-dimensional space represented as three 4×4 matrices etc..

The quaternion has a regular structure because it is a division algebra. When the appropriate quaternion variables are zero, that regular structure will rotate a 2-dimensional vector into a 2-dimensional vector; for example:

$$
\begin{bmatrix}
a & 0 & c & 0 \\
0 & a & 0 & c \\
-c & 0 & a & 0 \\
0 & -c & 0 & a
\end{bmatrix}
\begin{bmatrix}
\psi_1 \\ 0 \\ \psi_3 \\ 0
\end{bmatrix}
=
\begin{bmatrix}
a\psi_1 + c\psi_3 \\
0 \\
-c\psi_1 + a\psi_3 \\
0
\end{bmatrix}
\tag{22.36}
$$

or:

$$
\begin{bmatrix}
0 & b & c & 0 \\
-b & 0 & 0 & c \\
-c & 0 & 0 & -b \\
0 & -c & b & 0
\end{bmatrix}
\begin{bmatrix}
0 \\ \psi_2 \\ \psi_3 \\ 0
\end{bmatrix}
=
\begin{bmatrix}
b\psi_2 + c\psi_3 \\
0 \\
0 \\
-c\psi_2 + b\psi_3
\end{bmatrix}
\tag{22.37}
$$

We can do the same in 4-dimensions:

$$
\begin{bmatrix}
a & b & c & d \\
-b & a & -d & c \\
-c & d & a & -b \\
-d & -c & b & a
\end{bmatrix}
\begin{bmatrix}
\psi_1 \\ \psi_2 \\ \psi_3 \\ \psi_4
\end{bmatrix}
=
\begin{bmatrix}
a\psi_1 + b\psi_2 + c\psi_3 + d\psi_4 \\
-b\psi_1 + a\psi_2 - d\psi_3 + c\psi_4 \\
-c\psi_1 + d\psi_2 + a\psi_3 - b\psi_4 \\
-d\psi_1 - c\psi_2 + b\psi_3 + a\psi_4
\end{bmatrix}
\tag{22.38}
$$

However, we cannot do this with a 3-dimensional vector (we get a 4-dimensional vector from a 3-dimensional vector):

$$\begin{bmatrix} a & b & c & 0 \\ -b & a & 0 & c \\ -c & 0 & a & -b \\ 0 & -c & b & a \end{bmatrix} \begin{bmatrix} \psi_1 \\ \psi_2 \\ \psi_3 \\ 0 \end{bmatrix} = \begin{bmatrix} a\psi_1 + b\psi_2 + c\psi_3 \\ -b\psi_1 + a\psi_2 \\ -c\psi_1 + a\psi_3 \\ -c\psi_2 + b\psi_3 \end{bmatrix} \tag{22.39}$$

We see that we can do 2-dimensional and 4-dimensional $SU(2)$ rotations within the division algebra which is spinor space, but that we cannot do 3-dimensional rotations in spinor space. This coincides with the 2-dimensional and 4-dimensional representations of $SO(3)$ being spinor representations whereas the 3-dimensional representation is not a spinor representation. Conventionally, because $SU(2)$ is a double cover of $SO(3)$, the spinor representations of $SO(3)$ are thought of as being half-integral $\left\{ \dfrac{1}{2}, \dfrac{3}{2} \right\}$ and the non-spinor representations are thought of as being integral $\{1\}$.

Clifford algebras and $SO(n)$:

For every Clifford algebra with n basis vectors, it is possible to construct a n-dimensional representation of the Lie group $SO(n)$[136]. This representation is presented as a set of non-commuting matrices given by:

$$M_{jk} = \frac{1}{4i} \left[\overrightarrow{e_j}, \overrightarrow{e_k} \right] \tag{22.40}$$

For example, using $Cl_{3,0}$, we have:

---

[136] Georgi: page 270

$$M = \frac{1}{4i}\begin{bmatrix} \left[\vec{e_1},\vec{e_1}\right] & \left[\vec{e_1},\vec{e_2}\right] & \left[\vec{e_1},\vec{e_3}\right] \\ \left[\vec{e_2},\vec{e_1}\right] & \left[\vec{e_2},\vec{e_2}\right] & \left[\vec{e_2},\vec{e_3}\right] \\ \left[\vec{e_3},\vec{e_1}\right] & \left[\vec{e_3},\vec{e_2}\right] & \left[\vec{e_3},\vec{e_3}\right] \end{bmatrix}_j = \frac{2}{4i}\begin{bmatrix} 0 & \vec{e_{12}} & \vec{e_{13}} \\ \vec{e_{21}} & 0 & \vec{e_{23}} \\ \vec{e_{31}} & \vec{e_{32}} & 0 \end{bmatrix}$$

(22.41)

$$= \frac{1}{2i}\begin{bmatrix} 0 & \vec{e_{12}} & \vec{e_{13}} \\ -\vec{e_{12}} & 0 & \vec{e_{23}} \\ -\vec{e_{13}} & -\vec{e_{23}} & 0 \end{bmatrix}$$

We split the matrix into parts:

$$M = \frac{1}{2}\begin{bmatrix} 0 & -i & 0 \\ i & 0 & 0 \\ 0 & 0 & 0 \end{bmatrix}\vec{e_{12}} + \frac{1}{2}\begin{bmatrix} 0 & 0 & -i \\ 0 & 0 & 0 \\ i & 0 & 0 \end{bmatrix}\vec{e_{13}} + \frac{1}{2}\begin{bmatrix} 0 & 0 & 0 \\ 0 & 0 & -i \\ 0 & i & 0 \end{bmatrix}\vec{e_{23}}$$ (22.42)

We see that, although the main theme of this book is writing Clifford algebras as $C_2 \times C_2 \times \ldots$ division algebras, we can associate the $SO(n)$ Lie groups with each Clifford algebra. We thus associate the $SO(n)$ Lie groups with the rotation matrices of division algebras; the association in the quaternion case is a double cover and the association in the 8-dimensional case is as a double cover for six of the seven 2-dimensional sub-algebras (the e variable is not double covered), see (13.14).

Quaternions and $SO(3)$:

Suppose we use the imaginary quaternion variables as we have done above with the basis vectors of $Cl_{3,0}$. We have:

$$M = \frac{1}{2}\begin{bmatrix} [i,i] & [i,j] & [i,k] \\ [j,i] & [j,j] & [j,k] \\ [k,i] & [k,j] & [k,k] \end{bmatrix} = \begin{bmatrix} 0 & k & -j \\ -k & 0 & i \\ j & -i & 0 \end{bmatrix}$$

$$= \begin{bmatrix} 0 & k & 0 \\ -k & 0 & 0 \\ 0 & 0 & 0 \end{bmatrix} + \begin{bmatrix} 0 & 0 & -j \\ 0 & 0 & 0 \\ j & 0 & 0 \end{bmatrix} + \begin{bmatrix} 0 & 0 & 0 \\ 0 & 0 & i \\ 0 & -i & 0 \end{bmatrix}$$

(22.43)

Multiplying each of these by the appropriate imaginary variable and by a real variable will give three single variable anti-symmetric matrices whose exponentials are the three 2-dimensional Euclidean rotation matrices associated with the spatial part of our space-time. Is this anything more than a neat trick?

Chapter 23

Pure Quaternion Rotation

A pure quaternion is a quaternion with a zero real variable. We will denote a pure quaternion by:

$$\hat{r} = 0 + ix + jy + kz \tag{23.1}$$

It is a remarkable fact that, for any non-zero quaternion, $Q \in \mathbb{H}$, the expression $Q \, \hat{r} \, Q^{-1}$ is also a pure quaternion with the same length as $\hat{r}$ [137]. Of course the pure quaternions are of the same length because $Det(Q^{-1}) = (Det(Q))^{-1}$ and length is the determinant:

$$\left| Q \, \hat{r} \, Q^{-1} \right| = \left| \hat{r} \right| \tag{23.2}$$

The pure quaternions are taken to be a representation of the 3-dimensional spatial rotations of $SO(3)$. (Are three imaginary variables really representative of rotation in $\mathbb{R}^3$? I think not, but this is the standard mantra.) We can multiply both the Q matrices by minus unity. This is no more than multiplying the whole expression by unity. We have:

$$Q \, \hat{r} \, Q^{-1} = (-Q) \, \hat{r} \, \left(-Q^{-1}\right) \tag{23.3}$$

And so, within the quaternions, we have two 3-dimensional rotations that give the same result. This is a double cover; we say $SU(2)$ is a double cover of $SO(3)$; we mean that the quaternions are a double cover of the 3-dimensional spatial rotations within our space-time. If

---

[137] Lounesto: Page 70

we accept the definition of spinors as being elements of a double cover of $SO(n)$ groups, quaternions are spinors.

There is nothing remarkable other than that $Q\,\hat{r}\,Q^{-1}$ preserves the pure nature of $\hat{r}$. In a commutative division algebra, the Q and its inverse would cancel. In a non-commutative algebra, it seems that the Q and its inverse seems to 'kind-of-cancel' regarding the pure nature of the $\hat{r}$. Within a division algebra, the inverse of an element of that algebra is the conjugate of that element except for multiplication, or division, by the determinant. In polar form, the conjugate is the reverse rotation.

If we rotate the pure part of a 4-dimensional A_3 division algebra or the pure part (all seven imaginary variables) of an 8-dimensional division algebra using the $Q\,\hat{r}\,Q^{-1}$ type of rotation, we find a similar preservation of purity. It seems from calculation that this is the case with all division algebras, but your author does not properly understand why this is so nor is he able to prove this.

This 'preservation of purity' property of quaternions is often seen as being at the heart of the double cover of $SO(3)$ by $SU(2)$. In the case of the A_3 algebras, we get a double cover of $SO(1,2)$. The 8-dimensional division algebras have seven imaginary variables, and so have seven 2-dimensional sub-algebras, but, as we have said before, only six of these are double covers.

Chapter 24

Quaternions and two component Spinors

We follow Landau & Lifshitz in this next few paragraphs[138]. The wave function of a particle with spin $\frac{1}{2}$ has two components:

$$\psi = \begin{bmatrix} \psi^1 \\ \psi^2 \end{bmatrix} = \begin{bmatrix} \psi\left(\frac{1}{2}\right) \\ \psi\left(-\frac{1}{2}\right) \end{bmatrix} = \begin{bmatrix} a+ib \\ c+id \end{bmatrix} \tag{24.1}$$

Which we call a spinor. The probability of finding the particle corresponding to the spinor (24.1) at a given point in space is determined by the real number:

$$P \sim \psi^1\psi^{1*} + \psi^2\psi^{2*} = (a+ib)(a-ib)+(c+id)(c-id) \tag{24.2}$$
$$= a^2 + b^2 + c^2 + d^2$$

Of course, the product of a quaternion and its conjugate is:

$$\begin{bmatrix} a & -b & -c & -d \\ b & a & d & -c \\ c & -d & a & b \\ d & c & -b & a \end{bmatrix}\begin{bmatrix} a & b & c & d \\ -b & a & -d & c \\ -c & d & a & -b \\ -d & -c & b & a \end{bmatrix} \tag{24.3}$$

---

[138] Landau & Lifshitz: Quantum Mechanics 3rd edition. Page 206.
ISBN: 0750635398

$$= \begin{bmatrix} a^2 + b^2 + c^2 + d^2 & 0 & 0 & 0 \\ 0 & \sim & 0 & 0 \\ 0 & 0 & \sim & 0 \\ 0 & 0 & 0 & \sim \end{bmatrix} \tag{24.4}$$

Already, we wonder if we can replace such two component spinors as (24.1) with quaternions.

Under any rotation of the co-ordinate system, the components of a spinor transform linearly:

$$\psi^{1'} = p\psi^1 + q\psi^2 \qquad : \qquad \psi^{2'} = r\psi^1 + s\psi^2$$

$$\begin{bmatrix} p & q \\ r & s \end{bmatrix} \begin{bmatrix} \psi^1 \\ \psi^2 \end{bmatrix} = \begin{bmatrix} p\psi^1 + q\psi^2 \\ r\psi^1 + s\psi^2 \end{bmatrix} \tag{24.5}$$

In general, $\{p, q, r, s\}$ are functions of the rotation angle. Consider the bi-linear form:

$$\psi^1 \phi^2 - \phi^1 \psi^2 \tag{24.6}$$

Rotating both individual spinors, $\left\{ \begin{bmatrix} \psi^1 \\ \psi^2 \end{bmatrix}, \begin{bmatrix} \phi^1 \\ \phi^2 \end{bmatrix} \right\}$ by left multiplication

by the $\{p, q, r, s\}$ rotation matrix and recombining the rotated spinors into the bi-linear form produces:

$$\left(\psi^1 \phi^2 - \phi^1 \psi^2 \right)' = (ps - qr)\left(\psi^1 \phi^2 - \phi^1 \psi^2 \right) \tag{24.7}$$

We see that the bi-linear form (24.6) rotates into itself other than multiplication by $(ps - qr)$ which is the determinant of the rotation matrix. It rotates as a scalar, and, as such, we must have:

$$(ps - qr) = 1 \tag{24.8}$$

We have a unitary transformation. Following conventional wisdom, this means we have:

$$\begin{bmatrix} p & q \\ r & s \end{bmatrix}^{\dagger} = \begin{bmatrix} p & q \\ r & s \end{bmatrix}^{-1} \quad \& \quad \begin{bmatrix} p^* & r^* \\ q^* & s^* \end{bmatrix} = \begin{bmatrix} s & -q \\ -r & p \end{bmatrix} \qquad (24.9)$$

$$p = s^* \quad \& \quad q = -r^*$$

Thus, our rotation matrix is of the form:

$$\begin{bmatrix} w+ix & y+iz \\ -y+iz & w-ix \end{bmatrix} \equiv \begin{bmatrix} w & x & y & z \\ -x & w & -z & y \\ -y & z & w & -x \\ -z & -y & x & w \end{bmatrix} : Det = 1 \qquad (24.10)$$

In other words, we have shown that the matrix which rotates a two component spinor is a quaternion rotation matrix. Together, (24.2) & (24.9) imply that there are only three independent real parameters[139]. These are the three quaternion angles.

The argument that a two component spinor is really a quaternion is now becoming persuasive.

There are anti-quaternions, and so, presumably, there are anti-two component spinors. Presumably, this has implications regarding the Dirac equation.

---

[139] Landau & Lifshitz: Quantum Mechanics 3rd edition. Page 207. ISBN: 0750635398

Chapter 25

The Schrödinger Equation and Spinors

Conventionally, spinors are defined to be the elements of spin groups. Above, in our rewrite of Clifford algebra as the $C_2 \times C_2 \times ...$ division algebras, we have identified spin groups with the rotation matrices of division algebras. The Schrödinger equation, and the Dirac equation, contains spinors although in the case of the basic Schrödinger equation the single element spinor is often called a complex wave function. Since a spinor is an element of a division algebra, the Schrödinger equation, and the Dirac equation, ought to be written in the appropriate division algebra.

In this chapter, we follow Lounesto[140] for which we are most grateful.

The basic Schrödinger equation:
The basic Schrödinger equation presented to readers of introductory texts on quantum mechanics is:

$$ i\hbar \frac{\partial \psi}{\partial t} = -\frac{\hbar^2}{2m} \nabla^2 \psi + W\psi \qquad (25.1) $$

The basic Schrödinger equation is a non-relativistic quantum mechanical description of the 'spinless' electron in the absence of magnetic fields; by 'spinless' we mean the equation takes no account of the electron's intrinsic spin. The wave function, ψ, is single complex number, $\psi \in \mathbb{C}$, that is a single component spinor which is a function of the four space-time co-ordinates; this means it has a particular value at each point in space-time. We have:

---

[140] Pertti Lounesto: Chapter 4

$$\psi = \begin{bmatrix} f(t,x,y,z) & g(t,x,y,z) \\ -g(t,x,y,z) & f(t,x,y,z) \end{bmatrix} \qquad (25.2)$$

The Schrödinger equation in an electromagnetic field:
The basic Schrödinger equation is only a part of the full Schrödinger equation. We have little interest in only the basic Schrödinger equation but are concerned with the complete Schrödinger equation.

In an electromagnetic field with electric field $\vec{E}$ derived from the scalar potential V_{Pot} and magnetic field $\vec{B}$ derived from the vector potential $\vec{A}_{Pot}$, but still taking no account of the spin of the electron, the Schrödinger equation is[141]:

$$i\hbar\frac{\partial\psi}{\partial t} = \frac{1}{2m}\begin{bmatrix} -\hbar^2\nabla^2 + e^2\left(\overrightarrow{A_{Pot}}\bullet\overrightarrow{A_{Pot}}\right) \\ +i\hbar e\left(\nabla\bullet\overrightarrow{A_{Pot}} + \overrightarrow{A_{Pot}}\bullet\nabla\right) \end{bmatrix}\psi - eV_{Pot}\psi$$

$$= \frac{1}{2m}\left[\overrightarrow{\pi}^2\right]\psi - eV_{Pot}\psi \qquad (25.3)$$

Within this equation, (25.3), $\psi(t,x,y,z)$ is still a single complex number, and $\vec{\pi} = \vec{p} - e\vec{A}$. e is the charge of the electron.

The Schrödinger equation with electron spin:
Using the notation $\vec{\pi}$ to mean:

$$\pi_x = -i\hbar\frac{\partial}{\partial x} - eA_x \qquad \pi_y = -i\hbar\frac{\partial}{\partial y} - eA_y$$

$$\pi_z = -i\hbar\frac{\partial}{\partial z} - eA_z \qquad (25.4)$$

---

[141] Lounesto: Page 51

Using $\vec{p} = -i\hbar\nabla$, and remembering that operators do not commute, leads to:

$$\vec{\pi}^2 = \pi\bullet\pi = \left(p_x - eA_x\right)^2 + \left(p_y - eA_y\right)^2 + \left(p_z - eA_z\right)^2$$

$$= \vec{p}\bullet\vec{p} + e^2\vec{A}\bullet\vec{A} - e\left(\vec{p}\bullet\vec{A} + \vec{A}\bullet\vec{p}\right) \tag{25.5}$$

$$= -\hbar^2\nabla^2 + e^2\left(\overrightarrow{A_{Pot}}\bullet\overrightarrow{A_{Pot}}\right) + i\hbar e\left(\nabla\bullet\overrightarrow{A_{Pot}} + \overrightarrow{A_{Pot}}\bullet\nabla\right)$$

Using the notation $\vec{\sigma}$ to mean the Pauli matrices:

$$\sigma_x = \begin{bmatrix} 0 & 1 \\ 1 & 0 \end{bmatrix}, \quad \sigma_y = \begin{bmatrix} 0 & -i \\ i & 0 \end{bmatrix}, \quad \sigma_z = \begin{bmatrix} 1 & 0 \\ 0 & -1 \end{bmatrix} \tag{25.6}$$

We obtain the complete, non-relativistic, Schrödinger equation, also known as the Schrödinger-Pauli equation or just the Pauli equation, by replacing $\vec{\pi}^2$ in equation (25.3) by $\left(\vec{\sigma}\bullet\vec{\pi}\right)^2$:

$$i\hbar\frac{\partial\psi}{\partial t} = \frac{1}{2m}\left[\begin{array}{l} -\hbar^2\nabla^2 + e^2\left(\overrightarrow{A_{Pot}}\bullet\overrightarrow{A_{Pot}}\right) \\ +i\hbar e\left(\nabla\bullet\overrightarrow{A_{Pot}} + \overrightarrow{A_{Pot}}\bullet\nabla\right) - e\hbar\left(\vec{\sigma}\bullet\vec{B}\right) \end{array}\right]\psi - eV_{Pot}\psi \tag{25.7}$$

$$= \frac{1}{2m}\left[\vec{\pi}^2 - e\hbar\left(\vec{\sigma}\bullet\vec{B}\right)\right]\psi - eV_{Pot}\psi$$

This was first formulated by Wolfgang Pauli in 1927[142]; it is sometimes presented as:

$$\left[\frac{1}{2m}\left(\vec{\sigma}\bullet\left(\vec{p} - q\vec{A}\right)\right)^2 + qV\right]|\psi\rangle = i\hbar\frac{\partial}{\partial t}|\psi\rangle \tag{25.8}$$

Within this equation, (25.7) or (25.8) the spinor is a two component object (two complex numbers):

---

[142] Wolfgang Pauli (1927) Zur Quantenmachanik des magnetischen Elektrons Zeitschrift fur physic (43) 601-623

$$|\psi\rangle = \begin{bmatrix} \psi_+ \\ \psi_- \end{bmatrix} \qquad (25.9)$$

This much is standard Clifford algebra[143]. The spin term is:

$$\frac{e\hbar}{2m}\left(\vec{\sigma}\cdot\vec{B}\right) \qquad (25.10)$$

Wherein, we have:

$$\vec{\sigma}\cdot\vec{B} = \begin{bmatrix} 0 & 1 \\ 1 & 0 \end{bmatrix}\begin{bmatrix} B_x & 0 \\ 0 & B_x \end{bmatrix} + \begin{bmatrix} 0 & -i \\ i & 0 \end{bmatrix}\begin{bmatrix} B_y & 0 \\ 0 & B_y \end{bmatrix} + \begin{bmatrix} 1 & 0 \\ 0 & -1 \end{bmatrix}\begin{bmatrix} B_z & 0 \\ 0 & B_z \end{bmatrix}$$

$$= \begin{bmatrix} B_z & B_x - iB_y \\ B_x + iB_y & -B_z \end{bmatrix}$$

$$(25.11)$$

Aside: We note with interest that:

$$i\vec{\sigma}\cdot\vec{B} = \begin{bmatrix} iB_z & iB_x + B_y \\ iB_x - B_y & -iB_z \end{bmatrix} \equiv \begin{bmatrix} 0 & B_z & B_y & B_x \\ -B_z & 0 & -B_x & B_y \\ -B_y & B_x & 0 & -B_z \\ -B_x & -B_y & B_z & 0 \end{bmatrix} \in \mathbb{H}$$

$$(25.12)$$

We now have an equation in 2×2 matrices with complex elements. These matrices act upon a two component spinor. (We ignore the need to normalise at the moment.):

$$\psi = \begin{bmatrix} \psi_1 \\ \psi_2 \end{bmatrix} = \begin{bmatrix} a + ib \\ c + id \end{bmatrix} \qquad (25.13)$$

---

[143] Lounesto: Page 51

The Pauli equation directly from the Hamiltonian:

Alternatively, we can get to the Pauli equation by simply using the Hamiltonian. In classical theory, the Hamiltonian of a charged particle in an electromagnetic field is given by:

$$H = \frac{1}{2m}\left(\vec{p} - \frac{e}{c}\vec{A}\right)^2 + eV \qquad (25.14)$$

We include spin by adding in the energy $\left(-\vec{\mu}\cdot\vec{B}\right)$ for magnetic moment $\vec{\mu}$ in magnetic field $\vec{B}$. This, eventually, leads to:

$$H = \frac{1}{2m}\left(\vec{\sigma}\cdot\left(\vec{p} - q\vec{A}\right)\right)^2 + qV \qquad (25.15)$$

This is an equation conventionally written as 2×2 matrices.

A quaternion version of the Pauli equation:

We begin with (25.7):

$$i\hbar\frac{\partial\psi}{\partial t} = \frac{1}{2m}\left[\begin{array}{l} -\hbar^2\nabla^2 + e^2\left(\overrightarrow{A_{Pot}}\cdot\overrightarrow{A_{Pot}}\right) \\ +i\hbar e\left(\nabla\cdot\overrightarrow{A_{Pot}} + \overrightarrow{A_{Pot}}\cdot\nabla\right) - e\hbar\left(\vec{\sigma}\cdot\vec{B}\right) \end{array}\right]\psi - eV_{Pot}\psi \qquad (25.16)$$

We write the spinor as a quaternion:

$$\psi = \begin{bmatrix} a & b & c & d \\ -b & a & -d & c \\ -c & d & a & -b \\ -d & -c & b & a \end{bmatrix} \qquad (25.17)$$

We write the 4-potential as:

$$\mathbb{H}_{Pot} = \begin{bmatrix} V & A_x & A_y & A_z \\ -A_x & V & -A_z & A_y \\ -A_y & A_z & V & -A_x \\ -A_z & -A_y & A_x & V \end{bmatrix} \tag{25.18}$$

The term $\left(\nabla \cdot \overrightarrow{A_{Pot}} + \overrightarrow{A_{Pot}} \cdot \nabla \right)$ is twice the E field resulting from the non-commutative differentiation of the potential with zero V.

$$\left(\nabla \cdot \overrightarrow{A_{Pot}} + \overrightarrow{A_{Pot}} \cdot \nabla \right)_{[1,1]} = E_{[1,1]} = \frac{\partial V}{\partial t} + \frac{\partial A_x}{\partial x} + \frac{\partial A_y}{\partial y} + \frac{\partial A_z}{\partial z}$$

$$\left(\nabla \cdot \overrightarrow{A_{Pot}} + \overrightarrow{A_{Pot}} \cdot \nabla \right)_{[1,2]} = E_{[1,2]} = \frac{\partial A_x}{\partial t} - \frac{\partial V}{\partial x}$$

$$\left(\nabla \cdot \overrightarrow{A_{Pot}} + \overrightarrow{A_{Pot}} \cdot \nabla \right)_{[1,3]} = E_{[1,3]} = \frac{\partial A_y}{\partial t} - \frac{\partial V}{\partial y} \tag{25.19}$$

$$\left(\nabla \cdot \overrightarrow{A_{Pot}} + \overrightarrow{A_{Pot}} \cdot \nabla \right)_{[1,4]} = E_{[1,4]} = \frac{\partial A_z}{\partial t} - \frac{\partial V}{\partial z}$$

The term $\left(\overrightarrow{A_{Pot}} \cdot \overrightarrow{A_{Pot}} \right)$ is simply the dot product of the potential with zero V.

$$\left(\overrightarrow{A_{Pot}} \cdot \overrightarrow{A_{Pot}} \right) = \begin{bmatrix} V^2 + A_x^2 + A_y^2 + A_z^2 & 0 & 0 & 0 \\ 0 & \sim & 0 & 0 \\ 0 & 0 & \sim & 0 \\ 0 & 0 & 0 & \sim \end{bmatrix} \tag{25.20}$$

We take the term $\vec{\sigma} \cdot \vec{B}$ to simply be the magnetic field, the B field that is derived from the potential by non-commutative differentiation.

$$\vec{\sigma} \cdot \vec{B}_{[1,1]} = 0$$

$$\vec{\sigma} \cdot \vec{B}_{[1,2]} = \frac{\partial A_y}{\partial z} - \frac{\partial A_z}{\partial y} \tag{25.21}$$

$$\vec{\sigma} \cdot \vec{B}_{[1,3]} = \frac{\partial A_z}{\partial x} - \frac{\partial A_x}{\partial z}$$

$$\vec{\sigma} \cdot \vec{B}_{[1,4]} = \frac{\partial A_x}{\partial y} - \frac{\partial A_y}{\partial x}$$

$$(25.22)$$

We are unsure about this, but have we rewritten the Pauli-Schrödinger equation as a quaternion equation?

Chapter 26

The Spinor Operator and $Cl_{3,0}$

Within conventional Clifford algebra, there is the concept of the spinor operator. We consider this concept. Following Lounesto[144], we take the matrix:

$$\Psi = \begin{bmatrix} \psi_1 & -\psi_2^* \\ \psi_2 & \psi_1^* \end{bmatrix} = \begin{bmatrix} a+ib & -c+id \\ c+id & a-ib \end{bmatrix} = \begin{bmatrix} a & b & -c & d \\ -b & a & -d & -c \\ c & d & a & -b \\ -d & c & b & a \end{bmatrix} \in \mathbb{H}$$

(26.1)

This is a quaternion (the c-variable could change sign) and an element of the even sub-algebra $Cl_{3,0}^+$. Within $Cl_{3,0}^+$ the spin of the electron is computed as:

$$\vec{s} = \Psi \vec{e_3} \Psi$$

(26.2)

In this, we can see the Ψ as an operator, and so Lounesto refers to this as the spinor operator[145]. Using this spinor operator, the Schrödinger-Pauli equation becomes:

$$i\hbar \frac{\partial \Psi}{\partial t} = \frac{1}{2m} \vec{\pi}^2 \Psi - \frac{e\hbar}{2m} \vec{B}\Psi\vec{e_3} - eV_{Pot}\Psi$$

(26.3)

This looks very much like a quaternion equation, but we note that $\vec{e_3}$ is not an element of the even sub-algebra $Cl_{3,0}^+$.

---

[144] Lounesto: Page 63

[145] Lounesto has another spinor operator on page 143 from $Cl_{1,3}^+$.

The spin calculation with the spinor operator:
Traditionally, the spin expectation values of the electron are calculated from the two component spinor, ψ, as:

$$s_i = \psi^\dagger \sigma_i \psi \tag{26.4}$$

Explicitly, this is:

$$s_1 = \begin{bmatrix} a-ib & c-id \end{bmatrix} \begin{bmatrix} 0 & 1 \\ 1 & 0 \end{bmatrix} \begin{bmatrix} a+ib \\ c+id \end{bmatrix} = 2(ac+bd)$$

$$s_2 = \begin{bmatrix} a-ib & c-id \end{bmatrix} \begin{bmatrix} 0 & -i \\ i & 0 \end{bmatrix} \begin{bmatrix} a+ib \\ c+id \end{bmatrix} = 2(ad-bc) \tag{26.5}$$

$$s_3 = \begin{bmatrix} a-ib & c-id \end{bmatrix} \begin{bmatrix} 1 & 0 \\ 0 & -1 \end{bmatrix} \begin{bmatrix} a+ib \\ c+id \end{bmatrix} = a^2 + b^2 - c^2 - d^2$$

We change the sign of the c variable in the Lounesto spinor operator. This is no more than the arbitrary way that we label the spin directions. We now calculate:

$$\Psi^* [b=1] \Psi \tag{26.6}$$

Where the $[b=1]$ is the quaternion b variable matrix with $b=1$, Ψ^* is a conjugate quaternion and Ψ is a quaternion. We get:

$$\begin{bmatrix} a & -b & -c & -d \\ b & a & d & -c \\ c & -d & a & b \\ d & c & -b & a \end{bmatrix} \begin{bmatrix} 0 & 1 & 0 & 0 \\ -1 & 0 & 0 & 0 \\ 0 & 0 & 0 & -1 \\ 0 & 0 & 1 & 0 \end{bmatrix} \begin{bmatrix} a & b & c & d \\ -b & a & -d & c \\ -c & d & a & -b \\ -d & -c & b & a \end{bmatrix}$$

$$= \begin{bmatrix} 0 & a^2+b^2-c^2-d^2 & 2(bc-ad) & 2(ac+bd) \\ \sim & \sim & \sim & \sim \\ \sim & \sim & \sim & \sim \\ \sim & \sim & \sim & \sim \end{bmatrix} \tag{26.7}$$

Other than the sign of the c position, we get the spin expectation values as required. The sign of the c position is arbitrary. If we use the other unit variable matrices, $c = 1$ or $d = 1$, we get similar results but from a different orientation.

We see that we can calculate the spin expectation values of the electron using nothing more than quaternions provided we allow that the spinor representing the electron is a quaternion rather than two complex numbers.

Chapter 27

The Dirac Equation and Gamma Matrices

The Dirac gamma matrices, γ^μ, are 4×4 matrices that satisfy the anti-commutator relations:

$$\{\gamma^\mu, \gamma^\nu\} = 2\eta^{\mu\nu} \tag{27.1}$$

where $\eta^{\mu\nu}$ is the Minkowski metric. This gives:

$$\left(\gamma^0\right) = 1, \quad \left(\gamma^1\right) = -1, \quad \left(\gamma^2\right) = -1, \quad \left(\gamma^3\right) = -1$$
$$\gamma^\mu \gamma^\nu = -\gamma^\nu \gamma^\mu \quad \forall \; \mu \neq \nu \tag{27.2}$$

Clearly, the γ^μ are a Clifford algebra. They are in fact the basis of the 16-dimensional Clifford algebra $Cl_{1,3} \equiv \{1, \; 5\sqrt{+1}, \; 10\sqrt{-1}\}$. The Dirac matrices appear prominently in the Dirac equation.

$$\left(i\gamma^\mu \partial_\mu - m\right)\psi = 0 \tag{27.3}$$

Wherein:

$$\psi = \begin{bmatrix} a+ib \\ c+id \\ e+if \\ g+ih \end{bmatrix} \tag{27.4}$$

This is considered to be a four component spinor. Hm! it should be an element of an 8-dimensional Clifford algebra, but it is not an element of an 8-dimensional Clifford algebra.

A questionable equation:

There is something nonsensical about the Dirac equation. The Dirac equation includes the 4×4 gamma matrices acting upon a 4-component spinor (four complex numbers). Such a spinor has eight independent components and really ought to be written as an 8×8 matrix, and so we have 4×4 matrices acting upon an 8×8 matrix. This is nonsense. We also act upon non-commutative entities with differentiation. Traditionally, 'normal' differentiation is used, but we really ought to use non-commutative differentiation.

There have been many attempts to write the Dirac equation as a Clifford algebra equation, but none are very satisfactory. Using the block multiplication properties of matrices, we could expand the gamma matrices to be 8×8 matrices, $Cl_{3,0}$, and this has been tried, but we then get a whole 'extra' set of commutation relations which we do not want, and, besides, the gamma matrices are elements of a 16-dimensional Clifford algebra and they ought to be written as 16×16 matrices. The conventional way of dealing with these problems is to select the even sub-algebra of the 8-dimensional algebra and to write this as 8×8 matrices. This even sub-algebra is a 4-dimensional algebra; it ought to be written as 4×4 matrices.

Furthermore, we know that the anti-quaternions play an essential role in electromagnetism representing anti-electromagnetism. The Dirac equation includes anti-electromagnetism. Surely, some anti-algebra like the anti-quaternions ought to be involved in the Dirac equation. In previous attempts to rewrite the Dirac equation as a Clifford algebra Clifford algebraists have not considered the anti-quaternions or any other anti-algebra because the traditional way of deriving Clifford algebras does not produce the anti-algebras.

A different approach:

The commutation relation of the Dirac matrices, (27.2), are the commutation relations of the quaternions which are isomorphic to the Clifford algebra $Cl_{0,2}$. They are also the commutation relations of the anti-quaternions but reversed. They are not the commutation relations of any other 4-dimensional division algebra. We therefore can use either the quaternions or the anti-quaternions to rewrite the Dirac equation. We know that the quaternions are concerned with matter and that the anti-quaternions are concerned with anti-matter. We can use both the quaternions and the anti-quaternions to write two versions of the Dirac equation.

There is an $i = \sqrt{-1}$ in the Dirac equation. We presume this ought to be either $\{\hat{i}, j, k\}$ when applied to the different gamma matrices and that it does nothing more than convert a real (homothetic) matrix into the appropriate imaginary matrix.

There are differentiation operators in the Dirac equation. Within the non-commutative division algebras, and within Clifford algebras, we use non-commutative differentiation. The use of non-commutative differentiation means that we have two fields, the E field and the B field, for each equation. The gamma matrices combine with the real differentiation matrix to make the $SU(2)$ differentiation operator (see prerequisites):

$$i\gamma^\mu \partial_\mu \equiv \begin{bmatrix} \partial_t & \partial_x & \partial_y & \partial_z \\ -\partial_x & \partial_t & -\partial_z & \partial_y \\ -\partial_y & \partial_z & \partial_t & -\partial_x \\ -\partial_z & -\partial_y & \partial_x & \partial_t \end{bmatrix} \tag{27.5}$$

We take ψ to be a quaternion:

$$\psi = \begin{bmatrix} \phi & A_x & A_y & A_z \\ -A_x & \phi & -A_z & A_y \\ -A_y & A_z & \phi & -A_x \\ -A_z & -A_y & A_x & \phi \end{bmatrix} \tag{27.6}$$

Thus, we have the two quaternion fields given by (we present only the top row elements of the quaternion matrices for presentational ease):

$$E_{[1,1]} = \frac{\partial \phi}{\partial t} + \frac{\partial A_x}{\partial x} + \frac{\partial A_y}{\partial y} + \frac{\partial A_z}{\partial z}$$

$$E_{[1,2]} = \left(\frac{\partial A_x}{\partial t} - \frac{\partial \phi}{\partial x} \right)$$

$$E_{[1,3]} = \left(\frac{\partial A_y}{\partial t} - \frac{\partial \phi}{\partial y} \right) \tag{27.7}$$

$$E_{[1,4]} = \left(\frac{\partial A_z}{\partial t} - \frac{\partial \phi}{\partial z} \right)$$

$$B_{[1,1]} = 0$$

$$B_{[1,2]} = \left(\frac{\partial A_y}{\partial z} - \frac{\partial A_z}{\partial y} \right)$$

$$B_{[1,3]} = \left(\frac{\partial A_z}{\partial x} - \frac{\partial A_x}{\partial z} \right) \tag{27.8}$$

$$B_{[1,4]} = \left(\frac{\partial A_x}{\partial y} - \frac{\partial A_y}{\partial x} \right)$$

The Dirac equation now becomes two quaternion equations, one for the E field and one for the B field. We write them each as four separate equations for presentational ease. The E field is:

$$\frac{\partial \phi}{\partial t} + \frac{\partial A_x}{\partial x} + \frac{\partial A_y}{\partial y} + \frac{\partial A_z}{\partial z} - m\phi = 0$$

$$\left(\frac{\partial A_x}{\partial t} - \frac{\partial \phi}{\partial x} \right) - mA_x = 0$$

$$\left(\frac{\partial A_y}{\partial t} - \frac{\partial \phi}{\partial y} \right) - mA_y = 0 \tag{27.9}$$

$$\left(\frac{\partial A_z}{\partial t} - \frac{\partial \phi}{\partial z} \right) - mA_z = 0$$

And the B field is:

$$0 - m\phi = 0$$

$$\left(\frac{\partial A_y}{\partial z} - \frac{\partial A_z}{\partial y} \right) - mA_x = 0$$

$$\left(\frac{\partial A_z}{\partial x} - \frac{\partial A_x}{\partial z} \right) - mA_y = 0 \qquad (27.10)$$

$$\left(\frac{\partial A_x}{\partial y} - \frac{\partial A_y}{\partial x} \right) - mA_z = 0$$

Hm! the B field seems to be massless and therefore ($m = 0$) without magnetic field. If it has no magnetic field, the B field is presumably without electric charge. It looks like a neutrino[146]. We do not know what we have here. Have we two particles corresponding to two different masses one of which is zero. If we allow two such particles, since the E field and the B field are tied together, putting the two masses into both sets of the equations seems to give us an electron and a neutrino.

The anti-quaternions Dirac equation:
If we do with the anti-quaternions what we have done above with the quaternions, we will get the same E field and a reversed B field. Perhaps the anti-quaternion Dirac equation differs from the quaternion Dirac equation by a sign. We propose that the anti-quaternion Dirac equation is the conjugate Dirac equation:

$$\left(-i\gamma^\mu \partial_\mu - m \right)\psi = 0$$

$$\left(i\gamma^\mu \partial_\mu + m \right)\psi = 0 \qquad (27.11)$$

Well! There we have it. It is unknown how successful the above quaternion and anti-quaternion forms of the Dirac equation are, and they might be nonsense. However, they are 'properly formulated' equations in that they are each written within a division algebra. We

---

[146] See : Dennis Morris : The Quaternion Dirac Equation. This book was published 18 months after 'The Naked Spinor'

suffer from not understanding from where the mass term derives; as with the original Dirac equation, we have put it in by hand rather than derived it.

There is something we do not understand about the Dirac equation. The traditional Dirac equation has had enormous success in particle physics, and yet it simply does not fit together properly. As we said above, we have 4×4 matrices acting upon what should be 8×8 matrices, and the 4×4 matrices sould really be 16×16 matrices. More research is needed in this area.

Chapter 28

Concluding Remarks

I believe we can claim to have completely rewritten Clifford algebra as the $C_2 \times C_2 \times ...$ division algebras. The rewrite has tidied Clifford algebra considerably. We no longer have multiple definitions or descriptions of spinors, and we understand double cover of the 2-dimensional rotations of the $SO(n)$ groups. We can differentiate non-commutatively in a way that was convoluted and questionable in the conventional Clifford algebra. Much more than this, we have found that there are two types of empty space, the spinor spaces (division algebra spaces) and the $\mathbb{R}^n$ spaces such as our 4-dimensional space-time. We have generalised the definition of spinors to include unit length elements of all division algebras, but, within this generalisation, we can still pick out the double cover spinors within the $C_2 \times C_2 \times ...$ division algebras as a distinct type of spinor.

We know how general relativity and our 4-dimensional space-time emerge from these spinor spaces together with classical electromagnetism.

We have lost something. We have lost the geometric interpretation of Clifford algebra that considered multi-vectors like bi-vectors to be objects different from vectors. We have replaced this geometric interpretation with the geometric structure of spinor spaces and the higher dimensional trigonometric functions.

There is still much to be done. We do not understand clearly how the 8-dimensional $C_2 \times C_2 \times ...$ spaces fold into 4-dimensions, and we do not have a general 'formula' for higher dimensional trigonometric functions but have to calculate them each time. We have attempted to rewrite the Dirac equation, but it is not known if this is correct, and we do not really know what we are doing. Similarly, we have not even looked at the color forces which we think are within the 8-dimensional $C_2 \times C_2 \times ...$ spaces. None-the-less, in stripping spinors back to their most naked form, we believe we have cleared a lot of smoke and mirrors from both mathematics and physics.

At the very start of this book we quoted Michael Atiyah speaking of spinors. Within that quotation are the words *"they describe the "square root" of geometry"*. We have seen that the double cover derives from a square root in the trigonometric functions and so double cover spinors do in this sense *"describe the "square root" of geometry"*. Thank you for the clue Michael.

Dennis Morris

Port Mulgrave

January 2015

Appendix 1

The 8-dimensional Matrix Form

When we calculate the scaling parameters of the 4-dimensional algebras, we begin with nine potential scaling parameters and, by insisting upon multiplicative closure, we are able to eliminate all but three of them leaving only the three actual scaling parameters. In every case except one, insisting upon multiplicative closure leads to the elimination of potential scaling parameters by a linear equation (with only one possible solution). In the exceptional case, we are led to a quadratic elimination equation with two solutions. Choosing the positive solution gives the commutative $\{A_1, A_2\}$ algebras. Choosing the negative solution gives the non-commutative algebras $\{A_3, \mathbb{H}\}$. By choice, because it is neat, we eliminate the $P_{4,1}$ (the parameter associated with the bottom left-hand corner element) potential scaling parameter with the quadratic elimination equation. With the negative root of $P_{4,1}$, The 8-dimensional algebraic matrix form is:

$$\begin{bmatrix} A & B \\ C & D \end{bmatrix} \tag{28.1}$$

Where:

$$A = \begin{bmatrix} a & b & c & d \\ b.P_{2,1} & a & d\dfrac{P_{2,1}}{P_{2,4}} & c.P_{2,4} \\ c.P_{3,1} & -d\dfrac{P_{3,1}}{P_{2,4}} & a & -b.P_{2,4} \\ -d\dfrac{P_{2,1}P_{3,1}}{P_{2,4}^{\,2}} & c\dfrac{P_{3,1}}{P_{2,4}} & -b\dfrac{P_{2,1}}{P_{2,4}} & a \end{bmatrix} \tag{28.2}$$

$$B = \begin{bmatrix} e & f & g & h \\[1em] f\dfrac{P_{2,1}}{P_{2,6}} & e.P_{2,6} & h\dfrac{P_{2,1}}{P_{2,8}} & g.P_{2,8} \\[1.5em] g\dfrac{P_{3,1}}{P_{3,7}} & -h\dfrac{P_{2,6}P_{3,1}}{P_{2,8}P_{3,7}} & e.P_{3,7} & -f\dfrac{P_{2,8}P_{3,7}}{P_{2,6}} \\[1.5em] -h\dfrac{P_{2,1}P_{3,1}}{P_{2,4}P_{2,8}P_{3,7}} & g\dfrac{P_{2,6}P_{3,1}}{P_{2,4}P_{3,7}} & -f\dfrac{P_{2,1}P_{3,7}}{P_{2,4}P_{2,6}} & e\dfrac{P_{2,8}P_{3,7}}{P_{2,4}} \end{bmatrix} \qquad (28.3)$$

$$C = \begin{bmatrix} e.P_{5,1} & f\dfrac{P_{5,1}}{P_{2,6}} & g\dfrac{P_{5,1}}{P_{3,7}} & h\dfrac{P_{2,4}P_{5,1}}{P_{2,8}P_{3,7}} \\[1.5em] f\dfrac{P_{2,1}P_{5,1}}{P_{2,6}^{\,2}} & e\dfrac{P_{5,1}}{P_{2,6}} & h\dfrac{P_{2,1}P_{5,1}}{P_{2,6}P_{2,8}P_{3,7}} & g\dfrac{P_{2,4}P_{5,1}}{P_{2,6}P_{3,7}} \\[1.5em] g\dfrac{P_{3,1}P_{5,1}}{P_{3,7}^{\,2}} & -h\dfrac{P_{3,1}P_{5,1}}{P_{2,8}P_{3,7}^{\,2}} & e\dfrac{P_{5,1}}{P_{3,7}} & -f\dfrac{P_{2,4}P_{5,1}}{P_{2,6}P_{3,7}} \\[1.5em] -h\dfrac{P_{2,1}P_{3,1}P_{5,1}}{P_{2,8}^{\,2}P_{3,7}^{\,2}} & g\dfrac{P_{3,1}P_{5,1}}{P_{2,8}P_{3,7}^{\,2}} & -f\dfrac{P_{2,1}P_{5,1}}{P_{2,6}P_{2,8}P_{3,7}} & e\dfrac{P_{2,4}P_{5,1}}{P_{2,8}P_{3,7}} \end{bmatrix} \qquad (28.4)$$

$$D = \begin{bmatrix} a & b.P_{2,6} & cP_{3,7} & d\dfrac{P_{2,8}P_{3,7}}{P_{2,4}} \\[1.5em] b\dfrac{P_{2,1}}{P_{2,6}} & a & d\dfrac{P_{2,1}P_{3,7}}{P_{2,4}P_{2,6}} & c\dfrac{P_{2,8}P_{3,7}}{P_{2,6}} \\[1.5em] c\dfrac{P_{3,1}}{P_{3,7}} & -d\dfrac{P_{2,6}P_{3,1}}{P_{2,4}P_{3,7}} & a & -b.P_{2,8} \\[1.5em] -d\dfrac{P_{2,1}P_{3,1}}{P_{2,4}P_{2,8}P_{3,7}} & c\dfrac{P_{2,6}P_{3,1}}{P_{2,8}P_{3,7}} & -b\dfrac{P_{2,1}}{P_{2,8}} & a \end{bmatrix} \qquad (28.5)$$

By setting the different parameters to the various permutations of $\{-1,+1\}$, we get the matrix form of all the 8-dimensional algebras.

Other Books by Dennis Morris

The Naked Spinor – a Rewrite of Clifford Algebra

Spinors exist in Clifford algebras. In this book, we explore the nature of spinors. This book is an excellent introduction to Clifford algebra.

Complex Numbers The Higher Dimensional Forms – Spinor Algebra

In this book, we explore the higher dimensional forms of complex numbers. These higher dimensional forms are connected very closely to spinors.

Upon General Relativity

In this book, we see how 4-dimensional space-time, gravity, and electromagnetism emerge from the spinor algebras. This is an excellent and easy-paced introduction to general relativity.

From Where Comes the Universe

This is a guide for the lay-person to the physics of empty space.

Empty Space is Amazing Stuff – The Special Theory of Relativity

This book deduces the theory of special relativity from the finite groups. It gives a unique insight into the nature of the 2-dimensional space-time of special relativity.

The Nuts and Bolts of Quantum Mechanics

This is a gentle introduction to quantum mechanics for undergraduates.

Quaternions

This book pulls together the often separate properties of the quaternions. Non-commutative differentiation is covered as is non-commutative rotation and non-commutative inner products along with the quaternion trigonometric functions.

The Uniqueness of our Space-time

This book reports the finding that the only two geometric spaces within the finite groups are the two spaces that together form our universe. This is a startling finding. The nature of geometric space is explained alongside the nature of division algebra space, spinor space. This book is a catalogue of the higher dimensional complex numbers up to dimension fifteen.

Lie Groups and Lie Algebras

This book presents Lie theory from a diametrically different perspective to the usual presentation. This makes the subject much more intuitively obvious and easier to learn. Included is perhaps the clearest and simplest presentation of the true nature of the Lie group $SU(2)$ ever presented.

The Physics of Empty Space

This book presents a comprehensive understanding of empty space. The presence of 2-dimensional rotations in our 4-dimensional space-time is explained. Also included is a very gentle introduction to non-commutative differentiation. Classical electromagetism is deduced from the quaternions.

The Electron

This book presents the quantum field theory view of the electron and the neutrino. This view is radically different from the classical view of the electron presented in most schools and colleges. This book gives a very clear exposition of the Dirac equation including the quaternion rewrite of the Dirac equation. This is an excellent introduction to particle physics for students prior to university, during university and after university courses in physics.

The Quaternion Dirac Equation

This small book (only 40 pages) presents the quaternion form of the Dirac equation. The neutrino mass problem is solved and we gain an explanation of why neutrinos are left-chiral. Much of the material in this book is drawn from 'The Electron'; this material is presented concisely and inexpensively for students already familiar with QFT.

An Essay on the Nature of Space-time

This small and inexpensive volume presents a view of the nature of empty space without the detailed mathematics. The expanding universe and dark energy is discussed.

Elementary Calculus from an Advanced Standpoint

This book rewrite the calculus of the complex numbers in a way that covers all division algebras and makes all continuous complex functions differentiable and integrable. Non-commutative differentiation is covered. Gauge covariant differentiation is covered as is the covariant derivative of general relativity.

Even Mathematicians and Physicists make Mistakes

This book points out what seems to be several important errors of modern physics and modern mathematics. Errors like the misunderstanding of rotation, the failure to teach the higher dimensional complex numbers in most universities, and the mathematical inconsistency of the Dirac equation and some casual errors are discussed. These errors are set in their historical ircumstances and there is discussion about why they happened and the consequences of their happening. There is also an interesting chapter on the nature of mathematical proof within our society, and several famous proofs are discussed (without the details).

Finite Groups – A Simple Introduction

This book introduces the reader to finite group theory. Many introductory books on finite groups bury the reader in geometrical examples or in other types of groups and lose the central nature of a finite group. This book sticks firmly with the permutation nature of finite groups and elucidates that nature by the extensive use of permutation matrices. Permutation matrices simplify the subject considerably. This book is probably unique in its use of permutation matrices and therefore unique in its simplicity.

Non-commutative Differentiation and the Commutator

(The Search for the Fermion Content of the Universe)

This book develops the theory of non-commutative differentiation from the fundamentals of algebra. We see what an algebraic operation (addition, multiplication) really is, and we discover that the commutator is a third fundamental algebraic operation within some division algebras. This leads to the first part of the derivation of the fermion content of the universe.

Index

Z